The Theory of Remainders

Andrea Rothbart
Webster University

DOVER PUBLICATIONS, INC.
Mineola, New York

Bibliographical Note

This Dover edition, first published in 2005, is a corrected republication of the work originally published by Janson Publications, Dedham, Massachusetts, in 1995.

International Standard Book Number: 0-486-44256-X

Manufactured in the United States of America
Dover Publications, Inc., 31 East 2nd Street, Mineola, N.Y. 11501

To my husband, Leo Rodenborn
in memoriam

and to our grandchildren
April, Shelly, Sonia, Bianca,
Brian, and Erin

Contents

Preface

*"For Minkowski as for Hilbert, number theory was the most wonderful creation of the human spirit, equally a science and the greatest of arts."**
— Constance Reid

Ant says Hilbert and Minkowski were right. Ant thinks that there is no area of mathematics more enchanting than number theory. But that's how Ant talks. As for me, I really don't know. I hadn't studied that much math before Ant and I began our excursion through this book. Oh, I took high school algebra several years ago which I sort of remembered, but that's about it. Personally, my passions are magic and games. As a matter of fact, that's how Ant enticed me into this book: We saw a television show in which David Copperfield performed several tricks, including one where he used magic to catch a murderer on a train. Ant immediately saw how the trick was done and explained it to me. Copperfield was just using number theory! Then Ant showed me several other tricks using number theory. I realized then that if I was going to be a master magician I needed to participate in this book.

And now I'm beginning to see why Ant gets so wound up with the modular systems. You mess around with numbers, only the number systems are finite, and you solve equations, but it's in this really weird turf. And then, (and this is the best part), you use your knowledge to figure out things about football and chess! Incredible!

Gnam

**Source:* Constance Reid, *Hilbert* (New York: Springer-Verlag, 1970), 13.

I have long believed that mathematics, like music and literature and art, belongs to all the people. But that those of us who bask in the subject have, on the whole, done a wretched job of sharing our passion with others. I'm not exactly sure why. Perhaps we talk too quickly. Mathematics should be studied leisurely and playfully. One needs to "mess around" (Gnam's words). As you make your way through this book, remember there is no need to hurry. The subject you're about to encounter is incredibly meaty.

Andrea is too shy to appear in this book, but she did ask me to make a couple of comments on her behalf. She claims this book is appropriate for students who have studied (and, at least roughly understood) high school algebra. She has used it as a text with freshman liberal arts students, with sophomore mathematics majors (as an introduction to both number theory and abstract algebra), and with classes of high school and middle school teachers who were students in Webster University's Master of Arts in Teaching Mathematics Program.

Originally, Andrea wrote this manuscript because she wanted to provide middle school math teachers with a deeper understanding and broader perspective of arithmetic and algebra, and, equally important, with experience doing interesting mathematics. But her audience began expanding. To also accommodate those students with a little more mathematical background than Gnam, she included "problems" in addition to the exercises. The exercises provide experience with the concepts introduced and the opportunity to observe patterns, and are an integral part of this book. On the other hand, the problems are optional and should be chosen with discretion, depending upon the mathematical sophistication of the student.

Of course, all that is a pile of rot. I'm the principal author of this book. Not she!

Ant

Acknowledgments

"I think therefore I am."
— Rene Descartes, 1596 - 1650

I wish to express my gratitude to Andrea Rothbart and Gnam for their help with this book. However, rumors to the effect that I would have been unable to write it without their involvement are totally unfounded. Regarding the somewhat intrusive personal questions I have been asked, two of us are female and one of us is male. And, yes, two of us are fictitious and one of us is real. But for our purposes here, it is immaterial which of us are which. I hope these pages give you pleasure. But if not, as principal author I bear full responsibility, except for the notation used in analyzing Male Chauvinist Chess. Gnam is to blame for that.

Ant

"If I thank does that mean I am?"
— Gnam

I am indebted to Richard Singer for sharing with me the fruits of his creativity and his mathematical perspective. Many of his ideas have been incorporated into this book. And I am grateful to Richard for having read earlier versions of the manuscript, and providing many excellent suggestions for improvement. Richard has been my most cherished colleague for more than thirty years.

When good fortune was being distributed, I got more than my fair share in the person of Eric Karnowski, my editor at Janson Publications. He has been a joy to work with. Eric is supportive, capable, and always right-on-target with his recommendations. And he spent his weekends at home fixing my poorly drawn diagrams, all the while assuring me that he actually enjoyed doing it!

I have always considered myself the world's most spoiled teacher, for I have been consistently blessed with able and enthusiastic students. In particular, it has been my students who have kept me fueled with ideas and direction as this manuscript has undergone its numerous revisions. I remember, and am grateful to, each of my students for willingly assuming the role of guinea pig as this book was being developed, sharing ideas for improving it, and correcting my errors. (I call them typos!) In particular, I thank Don Love and Ev Corich.

My husband, Leo Rodenborn, was supportive and helpful in a variety of ways. He critiqued, he typed, and he never complained when I stayed up most of the night working and then slept through half the morning. Leo died shortly before this book first went to press.

Andrea

The Sections

Section One

The Black Magic Device

Ah, yes. Some of us haven't yet met. I'm Ant, an Amateur Number Theorist. Gnam, a Game Nut and Magician, ought to be along soon.

I must remember to tell Gnam about Simon Flagg. It seems that Simon had a rather extraordinary experience with the devil. After spending many years trying to either prove or disprove Fermat's Last Theorem*, Simon offered his soul to the devil in return for either a proof or a counter-example to the famous conjecture. If the devil couldn't settle the question within 24 hours, then Simon and his wife were assured of wealth, health and happiness for the rest of their days. Otherwise, well, so much for Simon's soul!

Apparently the devil was delighted with the deal (for he is adept at performing quite extraordinary feats) until he realized that it meant he had to do some mathematics. Mathematics belongs to the realm of the devil's nemesis, you know. Nevertheless, he set about his study in a serious manner. He even made trips to other planets to find out how extra-terrestrial beings were coping with the problem.

I think I hear Gnam coming now, so if you would like to know how the tale ends, you can read "The Devil and Simon Flagg" by Arthur Gorges in Clifton Fadiman's *Fantasia Mathematica.*

"Look at this strange device I have", said Gnam. The device looked like a piece of jewelry. It was round, and in the middle was a pentagonal black onyx prism. The edge was set in a lead

---

*Fermat's Last Theorem states that for any integer n greater than 2, there are no nonzero integers x, y, and z such that $x^n + y^n = z^n$. Later in this book, Fermat's Last Theorem will be discussed further.

ring. The inscriptions on the ring were all in indigo, whereas the inscriptions on the onyx were white. The symbols on the edge reminded me of those Gnam had once shown me in a medieval book on magic.

"A magician friend of mine says that she found this in a collection of black magic devices," continued Gnam. "She claims that it can do mathematics. But how can that be? Black magic is the realm of the devil, you know."

What a coincidence, I thought, as I proceeded to tell Gnam the story about Simon and the devil. Then I asked Gnam how to use the device.

"I am told that if you say **Oh Canine of Darkness here is my offering of 2000** that..."

Gnam never finished the sentence, for at that moment the device became hot, the onyx turned red, and the face of a black toy poodle with the name "Magic" written underneath seemed to glow in the center.

After composing myself, I bravely took the device from Gnam and began experimenting with it. I discovered that if I offered any integer between –1000 and 1000 to the dark canine, a number (usually different from the one offered) would appear in the center of the device. However, when I offered any other number, the face of the canine would appear.

In case you are interested, here is a list of the numbers I offered the device and the numbers that appeared in the center.

Number Offered Device	47	40	15	14	13	12	–24	130	–17
Number Appearing In Center	8	1	2	1	0	12	2	0	9

Naturally, I figured out rather quickly what the device was doing. (Have you?) Gnam was impressed, but the device was merely giving the remainder of the number offered when the number is divided by 13, and I am used to thinking in terms of remainders. Gnam didn't understand how the device worked with negative numbers; I promised to explain later.

I decided to experiment a little further with the device, to see if it could add or multiply. So I offered the number 14, and when the number 1 appeared in the center, I then offered "plus 15". Just as I expected, a 3, not a 2, now appeared in the center. "Plus 30" then elicited a 7. Next I offered "times 14" and the 7 remained. Then "times 15" and a 1 now appeared. Gnam was totally confused and asked me to make a list of some of the sequences of numbers I offered.

Number Offered Device	Number Appearing in Center
40	1
plus 53	2
plus 5	7
times 3	8
times 15	3
plus –2	1
plus –14	0

Number Offered Device	Number Appearing in Center
13	0
times –3	0
plus 41	2
times 24	9
times –12	9
plus –39	9
times 136	2
plus 265	7

After studying this list for several minutes, Gnam sighed. "The device is doing some kind of strange arithmetic, which I don't completely understand."

"We'll return to the question of what the device is doing, later. We'll even figure out how to use it to find the remainders of numbers that are larger than 1000 and numbers that are smaller than –1000. In Section 5, I will teach you techniques useful in solving problems such as..."

The Black Magic Device Problem

> Figure out a method (which employs no long division) of using the Black Magic Device to find the remainder of *any* integer, when divided by 13.

Some Comments on the Significance of Remainders

"It's very strange, Ant, that the Black Magic Device gives remainders but not quotients. (It's the ability to observe even such minor irregularities that separates us magicians from ordinary folk.)"

"Maybe not so strange. For it is remainders, far more than quotients, which provide insight into the behavior of numbers."

Gnam looked skeptical, so I added, "Elementary number theory deals with problems pertaining to the integers. Two integers can be added, subtracted or multiplied and the result is always an integer. But when you divide integers, you are likely to get a nonzero remainder."

Gnam was unimpressed, so I continued. "Most of the problems of number theory involve knowing when certain numbers are factors of other numbers. In fact, number theory might be characterized as the study of divisibility and factorization.

"The method of long division yields both the remainder and the quotient. When the remainder is zero, we know the divisor is a factor of the dividend. However, long division is a tedious way of checking for divisibility because most of the calculation is directed toward finding the quotient."

"How else would you find remainders?"

"You already know how to find remainders without using the long division process when dividing by 2 or 5."

"Or 10 or 20 or any of several other numbers, Ant. But those are rather special cases."

"Maybe not as special as you think. As I will show you later, when armed with a bit of insight into the theory of remainders, you can create remainder tests for divisors of your choice.

"The ancient Greeks discovered some general results pertaining to divisibility which were extended by mathematicians such as Pierre de Fermat and Leonhard Euler in the seventeenth and eighteenth centuries. However, in spite of the obvious connection between remainders and divisibility, the critical significance of remainders was not recognized until the nineteenth century when the mathematical giant Karl Friedrich Gauss used remainders as the central concept in his study of number theory. Instead of proceeding directly to the question of when is one number a factor of another number (that is, when is the remainder 0), Gauss built a theory of remainders by asking questions such as the following.

Given a fixed divisor:

1. When do two numbers have the same remainder?
2. What is the relationship between the remainders of $a, b, a + b, a - b$, and ab?
3. If a and b have the same remainder, and c and d have the same remainder, do $a + c$ and $b + d$ have the same remainder?

"By investigating these questions, we can develop a perspective that is very powerful in analyzing a variety of problems. The approach we will use is more modern than that devised by Gauss. Instead of studying remainders within the context of ordinary arithmetic, we will create new mathematical systems with new operations and study these. These systems are called **modular systems**. They are related to the integers, but are easier to study, largely because the modular systems are finite. Furthermore, our insights into the modular systems will enable us to solve a variety of interesting questions pertaining to the integers."

Section Two

The Modular Systems

"Back in Section 1, you promised to explain how the Black Magic Device handles negative numbers. Could it be that you conveniently forgot because you inconveniently don't know?"

"Unlike many nonmathematicians, I don't feel compelled to passively wait until other people explain things to me, Gnam. When an idea puzzles me, I play around with it until I understand. I question things! I suppose I've always known that if one doesn't want to carry around a lot of rubbish in one's head, it's necessary to be skeptical. Even as a child I questioned my teacher when she taught long division as follows:"

$$\begin{array}{r} 8 \\ 6\overline{)49} \\ 48 \\ \hline 1 \end{array} \qquad \begin{array}{r} 10 \\ 14\overline{)153} \\ 140 \\ \hline 13 \end{array} \qquad \begin{array}{r} 3 \\ 16\overline{)48} \\ 48 \\ \hline 0 \end{array} \qquad \begin{array}{r} 0 \\ 12\overline{)11} \\ 0 \\ \hline 11 \end{array}$$

"She checked her answers in the following way:"

$$49 = 8 \cdot 6 + 1, 153 = 10 \cdot 14 + 13, 48 = 3 \cdot 16 + 0, 11 = 0 \cdot 12 + 11.$$

"So I countered with:"

$$\begin{array}{r} 5 \\ 6\overline{)49} \\ 30 \\ \hline 19 \end{array}$$

"And as proof, I offered the check: $49 = 5 \cdot 6 + 19$."

"Did your teacher mark that wrong?" asked Gnam. "My teacher, Mr. Rob, was a Rigid Old (ahem) Being and *he* would have marked it wrong."

"No, Ms. Far was Flexible and Reasonable. She encouraged me to think through the implications of what would happen if we allowed this. I soon discovered that I could get any quotient I wanted. I could even get a quotient of 0 by letting the dividend itself be the remainder: $49 = 0 \cdot 6 + 49$. Then Ms. Far showed me how I could get a negative remainder."

$$\begin{array}{r} 9 \\ 6\overline{)49} \\ \underline{54} \\ -5 \end{array} \qquad \text{Check: } 49 = 9 \cdot 6 + -5.$$

"What are you doing, Ant? Remainders aren't supposed to be negative! Nor are they supposed to be larger than the divisor. Everybody knows that!"

"I agree, Gnam. The standard convention requires that the remainder be a nonnegative number less than the divisor. This insures that every long division problem has a unique quotient and a unique remainder."

The Division Algorithm

> If a and n are positive integers, then there exist unique integers q and r such that $a = qn + r$ where $0 \le r < n$.

Exercise 2.1

> Find the quotient q and the remainder r satisfying the Division Algorithm for each of the following pairs (a, n), where a is the dividend and n is the divisor.
>
> 1. $a = 35$ and $n = 18$
> 2. $a = 35$ and $n = 7$
> 3. $a = 35$ and $n = 40$

"My Friend, Mf, comes from a place called Ltac, where people Like to Alter Conventions. In particular, on the first Tuesday of every third month, the Ltacians agree to require all remainders to be greater than or equal to the divisor and less than twice the divisor. So, for example, when 14 is divided by 4, the quotient is 2 and the remainder is 6. The Ltacians arrive at this using repeated subtraction."

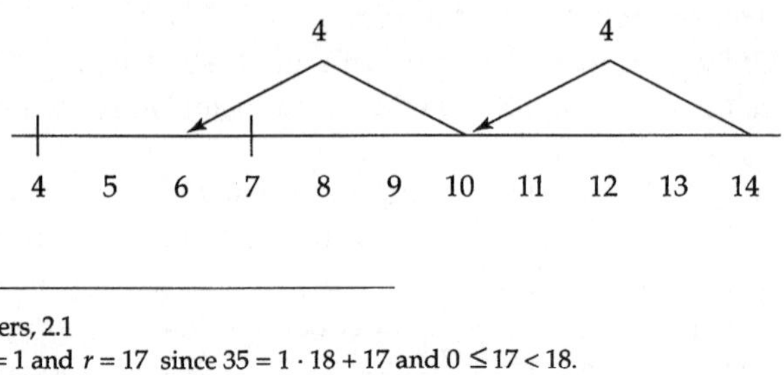

Answers, 2.1

1. $q = 1$ and $r = 17$ since $35 = 1 \cdot 18 + 17$ and $0 \le 17 < 18$.
2. $q = 5$ and $r = 0$ since $35 = 5 \cdot 7 + 0$ and $0 \le 0 < 7$.
3. $q = 0$ and $r = 35$ since $35 = 0 \cdot 40 + 35$ and $0 \le 35 < 40$.

$$\begin{array}{r} 2 \\ 4\overline{)14} \\ \underline{8} \\ 6 \end{array} \qquad \text{Check: } 14 = 2 \cdot 4 + 6 \text{ and } 4 \le 6 < 8.$$

"And on the second Wednesday of every fifth month, the remainder interval is the set of integers greater than or equal to the negative of the divisor but less than zero."

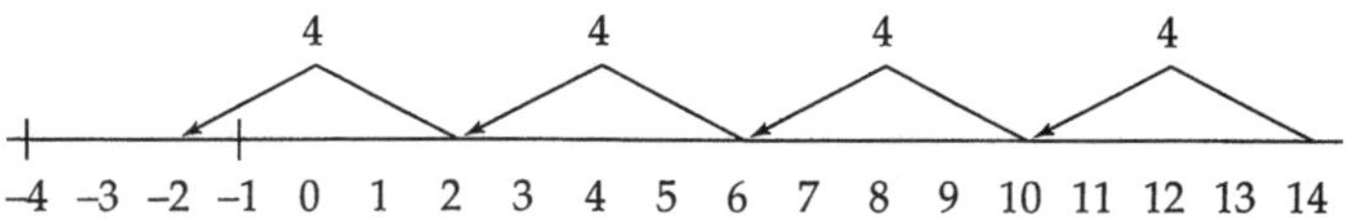

$$\begin{array}{r} 4 \\ 4\overline{)14} \\ \underline{16} \\ -2 \end{array} \qquad \text{Check: } 14 = 4 \cdot 4 + -2 \text{ and } -4 \le -2 < 0.$$

Exercise 2.2

1. How would the Ltacians do the long division problems in Exercise 2.1 on the first Tuesday of every third month when the remainder has to be greater than or equal to the divisor and less than twice the divisor?
2. How would the Ltacians do the problems in Exercise 2.1 on the second Wednesday of every fifth month when the remainder has to be greater than or equal to the negative of the divisor and less than 0?
3. What other intervals of integers would qualify as remainder intervals for division by 7?

Answers, 2.2

1. If $a = 35$ and $n = 18$ then the quotient is 0 and the remainder is 35 since $35 = 0 \cdot 18 + 35$ and $18 \le 35 < 36$.
   If $a = 35$ and $n = 7$ then the quotient is 4 and the remainder is 7 since $35 = 4 \cdot 7 + 7$ and $7 \le 7 < 14$.
   If $a = 35$ and $n = 40$ then the quotient is –1 and the remainder is 75 since $35 = -1 \cdot 40 + 75$ and $40 \le 75 < 80$.
2. $q = 2$ and $r = -1$ since $35 = 2 \cdot 18 + -1$; $q = 6$ and $r = -7$ since $35 = 6 \cdot 7 + -7$; $q = 1$ and $r = -5$ since $35 = 1 \cdot 40 + -5$.
3. Any set of seven consecutive integers would qualify. Some of the infinitely many possibilities for remainder intervals upon division by 7 are $\{70, 71, 72, 73, 74, 75, 76\}$, $\{-14, -13, -12, -11, -10, -9, -8\}$, and $\{3, 4, 5, 6, 7, 8, 9\}$. Indeed, if you're flexible enough to consider this, even the set $\{-14, -6, 2, 3, 18, 19, 27\}$ will do. You might want to think about why!

"I'm sure that you find this all quite fascinating, Ant. But you have been digressing again from your promise to explain how the device handles negative numbers."

"The point is that the Division Algorithm also holds if the dividend is negative. For example, consider the problem –35 divided by 15:"

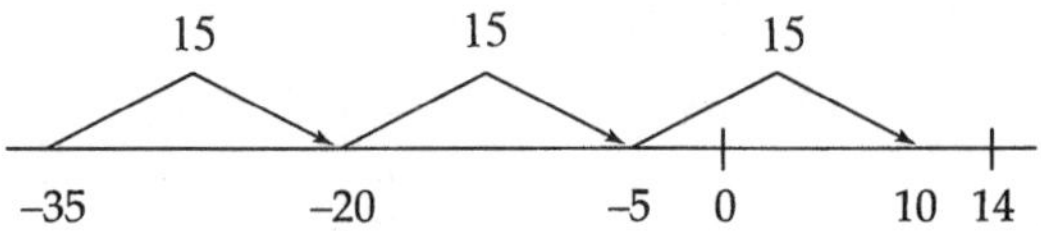

"If you subtract –3 fifteens from –35, you arrive at 10 which is in the conventional remainder interval."

$$\begin{array}{r} -3 \\ 15\overline{)-35} \\ \underline{-45} \\ 10 \end{array}$$

"Since $-35 = -3 \cdot 15 + 10$ and $0 \le 10 < 15$, the quotient is –3 and the remainder is 10."

Exercise 2.3

Find the quotient, q, and the remainder, r, that satisfy the equation $a = qn + r$ and $0 \le r < n$ for each of the following.

1. $a = -12$ and $n = 5$

2. $a = -12$ and $n = 13$

3. $a = -12$ and $n = 4$

4. $a = -1203$ and $n = 60$

Exercise 2.4

What remainder will the Black Magic Device give when –38 is entered? When –135 is entered? When –1 is entered? (Recall that the Black Magic Device computes remainders upon division by 13.)

Answers, 2.3

1. $q = -3$ and $r = 3$
2. $q = -1$ and $r = 1$
3. $q = -3$ and $r = 0$
4. $q = -21$ and $r = 57$

Answers, 2.4

1; 8; 12

Supplementary Problem 2.1*

How might the Division Algorithm be stated so that it applies to any integral divisor, except 0, and any integral dividend?

"Now that I understand how the Black Magic Device copes with negative inputs, let's move on to the question of how to use it with very large and very small numbers."

"Patience, Gnam! First I'd like us to investigate a device similar to the Black Magic Device, called the r_{11} machine. The r_{11} machine computes the remainder when an integer is divided by 11, but without going through the process of actually dividing by 11. Thus the r_{11} machine doesn't waste energy computing quotients too. But before I can show you how it works, I need to introduce some relevant notation and concepts.

"If an integer, a, is fed into the r_{11} machine, the output is denoted by $r_{11}(a)$. You can think of $r_{11}(a)$ as meaning 'what the machine r_{11} does to a'. For example, if you input the number 12, the notation '$r_{11}(12)$' represents the output. Since r_{11} computes remainders upon division by 11, $r_{11}(12) = 1$. (This is read as 'r sub eleven of twelve equals one.')"

Notation

'$\mathbf{r_{11}(a)}$' means 'the remainder when a is divided by 11'.

Examples: $r_{11}(25) = 3$; $r_{11}(5) = 5$; $r_{11}(-5) = 6$.

Exercise 2.5

1. What is $r_{11}(15)$? $r_{11}(-15)$? $r_{11}(3)$? $r_{11}(-3)$? $r_{11}(33)$? $r_{11}(-33)$? What is $r_{11}(46)$? What is my next question? What is its answer? In general, what is the relationship between $r_{11}(a)$and $r_{11}(-a)$?

***Note:** Throughout this book there are questions characterized as "problems". A problem may be a source of perplexity, requiring tenacity and self-confidence to pursue. It is the mathematician's lot to wrestle with conceptual problems without assurance of ultimately being able to solve any given one, but success, when it comes, is so satisfying. It is a wonderful feeling to experience the power of one's brain!

Depending upon one's mathematical sophistication, one person's problem may be another's exercise; however, I don't think you'll find this first problem very tough.

Unlike the exercises, completion of the problems is not assumed in the text. Answers to problems can be found at the end of this book.

2. What is $r_{11}(33)$? $r_{11}(-44)$? $r_{11}(0)$? Suppose that 11 is a factor of a. What can you conclude about $r_{11}(a)$?
3. What is $r_{11}(15)$? $r_{11}(26)$? What is $26 - 15$? What is $r_{11}(7)$? $r_{11}(-15)$? What is $7 - (-15)$? Suppose that $r_{11}(a) = r_{11}(b)$. What relationship exists between a and b?

Notation

$\mathbf{Z}_{11}$ denotes the set of remainders upon division by 11. That is, $\mathbf{Z}_{11} = \{0, 1, 2, 3, 4, 5, 6, 7, 8, 9, 10\}$.

"Gnam, machines with the property that they spew out something when they're fed something, and do so in a consistent manner (that is, for each acceptable input, there is a *unique* output) are called **functions**. r_{11} is an example of a function from the set of integers (the set of acceptable inputs) onto $\mathbf{Z}_{11}$ (the set of possible outputs)."

"This is incredibly interesting, Ant. Please don't let my yawning disturb you."

"The concept of a function is one of the basic concepts of mathematics. Functions are integral to nearly every area of mathematics."

"You can also ignore my snoring."

"Well wake up, Gnam, because I'm about to introduce an addition on the elements of $\mathbf{Z}_{11}$ called **mod 11 addition** and denoted by $+_{11}$. To motivate this definition, let's examine an 11-hour clock."

"An 11-hour clock?"

"Sure. Why not, Gnam? It's the kind of clock that is used on the moon Titan. It looks like this one.

Answers, 2.5

1. 4; 7; 3; 8; 0; 0; 2; What is $r_{11}(-46)$?; 9; $r_{11}(a) + r_{11}(-a) = 11$ except when $r_{11}(a)$ is 0. Then $r_{11}(a) + r_{11}(-a) = 0$.
2. 0; 0; 0; 11 is a factor of a if and only if $r_{11}(a) = 0$.
3. 4; 4; 11; 7; 7; 22; a and b have the same remainder if and only if their difference is a multiple of 11.

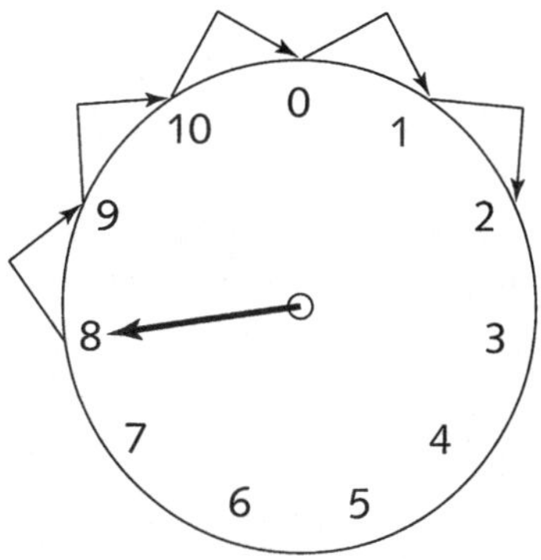

"Now, if it is 8 o'clock on Titan, and 5 moon hours elapse, then it becomes 2 o'clock. That is, 8 + 5 = 2 on an 11-hour clock. We could also express this idea as

$$8 +_{11} 5 = 2."$$

"I understand, Ant. But I don't see why the Titan's clocks have a 0 where the 11 belongs. Our 12-hour clock uses a 12, not a 0."

"There is a simple explanation for that," I responded. "Suppose that it is 3 o'clock on earth and 12 hours elapse. Then it's 3 o'clock again, right? In fact, any time t plus 12 is just t again on our clocks. On a 12–hour clock:

$$t + 12 = t \text{ and } 12 + t = t.$$

"In other words, what we label as 12 behaves just like zero. We should have labelled it 0 not 12! The Titans, with their superior intellect, understand this. Hence the 0, rather than an 11 on their 11-hour clocks."

Exercise 2.6

Complete the following mod 11 addition table.

$+_{11}$	0	1	2	3	4	5	6	7	8	9	10
0											
1											
2											
3											
4											
5											
6											
7											
8											
9											
10											

"The completed table constitutes a definition of mod 11 addition. But it is a rather bulky one, for it contains 121 distinct facts. It would be helpful to have a more concise definition."

"I thought of a way to add without using the clock or the table, Ant. A problem like $2 +_{11} 3$ is easy, since it has the same answer as in ordinary addition. But to find the answer to a problem such as $9 +_{11} 3$ just add like you usually do and then subtract 11. $9 + 3 = 12$, but you have to subtract 11, so $9 +_{11} 3 = 1$."

Gnam's Definition of mod 11 addition

> If a and b elements of $\mathbf{Z}_{11}$, then $a +_{11} b = a + b$ if $a + b < 11$, but $a +_{11} b = a + b - 11$ if $a + b \geq 11$.

"That works for mod 11 *addition*, Gnam, but..."

"You can even use my approach to define mod 11 *multiplication*! For example, $2 \cdot 3 = 6$ in both ordinary and mod 11 multiplication. But to find $2 \cdot_{11} 7$ just think 14 and then subtract 11 to get 3. (That's the correct answer since if you make two moves of 7 on an 11-hour clock, you land on 3.)"

"But the answer to $5 \cdot_{11} 9$ isn't $45 - 11 = 34$, Gnam. There is no 34 on the 11–hour clock. If, starting at 0 you move 9 five times (that is, move 45), you end up at 1, because 45 is 1 more than 44, a multiple of 11. 1 is the remainder when 45 is divided by 11. Or, as another example, to find $8 \cdot_{11} 5$ on an 11-hour clock, note that if you move 40, you make 3 cycles of 11 and then land on 7. That is, $8 \cdot_{11} 5 = r_{11}(8 \cdot 5) = r_{11}(40) = 7$. In general, to multiply (or add) two numbers mod 11, you can just multiply (or add) them regularly and then put the product (or sum) through the r_{11} machine."

Definition

> If a and b are elements of $\mathbf{Z}_{11}$, then $a +_{11} b = r_{11}(a + b)$ and $a \cdot_{11} b = r_{11}(ab)$.
>
> Examples: $8 \cdot_{11} 6 = r_{11}(8 \cdot 6) = r_{11}(48) = 4$.
> $9 \cdot_{11} 8 = r_{11}(9 \cdot 8) = r_{11}(72) = 6$.
> $5 +_{11} 9 = r_{11}(5 + 9) = r_{11}(14) = 3$.
> $2 +_{11} 6 = r_{11}(2 + 6) = r_{11}(8) = 8$.

Exercise 2.7

Complete the following mod 11 multiplication table.

$\cdot_{11}$	0	1	2	3	4	5	6	7	8	9	10
0											
1											
2											
3											
4											
5											
6											
7											
8											
9											
10											

Definition

"The set $\mathbf{Z}_{11}$ under the operations $+_{11}$ and $\cdot_{11}$ is called **the integers mod 11** and is an example of a modular system. *

Exercise 2.8

Compute the following. The usual arithmetic conventions hold (such as multiplication takes precedence over addition).

1. $4 \cdot_{11} 3 +_{11} 5$ 2. $8 +_{11} 3 +_{11} 9 +_{11} 2 +_{11} 6 +_{11} 5$
3. $5 \cdot_{11} 7 +_{11} 4 \cdot_{11} 7$ 4. $2 \cdot_{11} 7 +_{11} 7 \cdot_{11} 7$
5. $6 \cdot_{11} 9 \cdot_{11} 2$

"We're almost ready to develop a remainder test for 11. But we need the help of a very important idea. Suppose that the Black Magic Device gave remainders upon division by 11, rather than 13. So, if you offered it the number *21*, it would respond with *10*. If you then offered it *plus 12*, it would respond with *0* since $r_{11}(21) +_{11} r_{11}(12) = 10 +_{11} 1 = 0$.

Answers, 2.8

1. 6 2. 0 3. 8 4. 8 5. 9

*Sometimes this modular system is referred to as just $\mathbf{Z}_{11}$, when what is really meant is the set $\mathbf{Z}_{11}$ together with the operations $+_{11}$ and $\cdot_{11}$.

"But there is an alternate way of arriving at the answer 0. (Do you see it?) First add 21 to 12 getting 33, and then offer the device 33. That is,

$$r_{11}(21 + 12) = r_{11}(21) +_{11} r_{11}(12).$$

"Now consider $r_{11}(21 \cdot 12)$. If you multiply 12 by 21 you get 252 and $r_{11}(252) = 10$. But $r_{11}(21) \cdot_{11} r_{11}(12) = 10 \cdot_{11} 1$ which is also 10. In other words,

$$r_{11}(21 \cdot 12) = r_{11}(21) \cdot_{11} r_{11}(12).$$

"So if I want to find $r_{11}(21 \cdot 12)$, I can save myself a bit of fuss and bother by instead computing $r_{11}(21) \cdot_{11} r_{11}(12)$!"

Exercise 2.9

1. What is $r_{11}(12) +_{11} r_{11}(9)$? What is $r_{11}(12 + 9)$?
2. What is $r_{11}(15) \cdot_{11} r_{11}(23)$? What is $r_{11}(15 \cdot 23)$?
3. What is $r_{11}(110) +_{11} r_{11}(13)$? What is $r_{11}(110 + 13)$?
4. What is $r_{11}(79) \cdot_{11} r_{11}(25)$? What is $r_{11}(79 \cdot 25)$?
5. What is $[r_{11}(11) \cdot_{11} r_{11}(12)] +_{11} r_{11}(17)$?
   What is $r_{11}(11 \cdot 12 + 17)$?
6. Compute each of the following in two different ways, and decide which way (if either) is quicker.
   a. $r_{11}(11 \cdot 397 + [-6])$
   b. $r_{11}(12 + 13 + 14 + 15)$
   c. $r_{11}(443 \cdot 13)$
   d. $r_{11}(338 \cdot 552 + 994 \cdot 333)$

"Even though I don't know how it's going to help us figure out a remainder test for 11, Ant, I think I see what you're trying to say."

"The following theorem, a generalization of the ideas developed in the last exercise, is the key to developing remainder tests."

Answers, 2.9

1. 10, 10
2. 4, 4
3. 2, 2
4. 6, 6
5. 6, 6
6. a. $0 +_{11} 5 = 5$
   b. $1 +_{11} 2 +_{11} 3 +_{11} 4 = 10$
   c. $3 \cdot_{11} 2 = 6$
   d. $8 \cdot_{11} 2 +_{11} 4 \cdot_{11} 3 = 6$

Super-Important Theorem

If s and t are any two integers,

$$r_{11}(s+t) = r_{11}(s) +_{11} r_{11}(t) \text{ and}$$
$$r_{11}(s \cdot t) = r_{11}(s) \cdot_{11} r_{11}(t).$$

"Note that the Super-Important Theorem applies to *all* integers s and t, Gnam. When s and t are elements of $\mathbf{Z}_{11}$, $r_{11}(s) = s$ and $r_{11}(t) = t$. In that case, the theorem merely restates the definition of $+_{11}$ and $\cdot_{11}$."

Supplementary Problem 2.2*

1. Prove that for any integers a and b, $r_{11}(11a + b) = r_{11}(b)$. (Suggestion: Use the Division Algorithm.)
2. Prove the Super-Important Theorem.

Notation

Expressions such as

$$[r_{11}(16) +_{11} r_{11}(17)] \cdot_{11} r_{11}(31)$$

are tedious to read as well as to write. So frequently we will omit the subscript 11, and instead write **(mod 11)** to the right of expressions, equations, and sequences of equations to indicate that 11 is the subscript on all remainders and operations.

"For example, the above expression could be written as:

$$[r(16) + r(17)] \cdot r(31) \pmod{11}.$$

"Similarly, $8 + 4 = 1 \pmod{11}$ means $8 +_{11} 4 = 1$.

"I'm now ready to demonstrate a remainder test for 11. First, what is $r_{11}(100)$?"

***Note:** Some of the problems in this text call for proofs. Providing a proof entails first creating a compelling argument and then expressing it clearly. The latter is learned through experience and guidance. (I think I learned by reading other people's proofs and then trying to duplicate them.) The argument itself may be routine or, at the other extreme, a manifestation of great insight. Don't be alarmed if it is unclear to you how to proceed when you are asked to prove something. Almost everyone is confused at first (and at second and at third). With perseverance, you will catch on. When you do, you will have sharpened your ability to reason mathematically.

"That's easy, Ant. It's 1 since 11 is a factor of 99 and $100 = 99 + 1$."

"Okay, now $500 = 5 \cdot 100$, so what is $r_{11}(500)$?"

"Wait, I'll have to do some division. Just a minute now..."

"Use the Super-Important Theorem, Gnam. You know what $r_{11}(100)$ is and you know what $r_{11}(5)$ is, so what is $r_{11}(5 \cdot 100)$?"

"Oh, I see. It's 5:"

$$\begin{aligned} r(500) &= r(5) \cdot r(100) \pmod{11} \\ &= 5 \cdot 1 \pmod{11} \\ &= 5 \pmod{11}. \end{aligned}$$

Exercise 2.10

What is $r_{11}(600)$? $r_{11}(1000)$? $r_{11}(3400)$? In general, what is $r_{11}(k \cdot 100)$?

"Now consider a problem such as $r_{11}(3282)$."

$$3282 = 3200 + 82.$$

$$\begin{aligned} r(3282) &= r(32) \cdot r(100) + r(82) \pmod{11} \\ &= 10 \cdot 1 + 5 \pmod{11} \\ &= 4 \pmod{11}. \end{aligned}$$

"Similarly, $r(915) = r(9) \cdot r(100) + r(15) = 9 + 4 = 2 \pmod{11}$."

Exercise 2.11

Compute $r_{11}(638)$; $r_{11}(1259)$; $r_{11}(843)$; $r_{11}(4782)$.

"Now note that

$$r(10^2) = 1 \pmod{11}$$

so $$r(10^4) = r(10^2) \cdot r(10^2) = 1 \cdot 1 = 1 \pmod{11}$$

and $$r(10^6) = r(10^2) \cdot r(10^4) = 1 \cdot 1 = 1 \pmod{11}.$$

and so on. That is, For each $k \geq 0$, $r_{11}(10^{2k}) = 1$.

Answers, 2.10
6; 10; 1; $r_{11}(k)$.

Answers, 2.11
$6 +_{11} 5 = 0$; $1 +_{11} 4 = 5$; $8 +_{11} 10 = 7$; $3 +_{11} 5 = 8$.

"So to find $r_{11}(175460)$ one could reason as follows:

$$175460 = 17 \cdot 10^4 + 54 \cdot 10^2 + 60$$

$$\begin{aligned} \text{so} \quad r(175460) &= r(17) \cdot r(10^4) + r(54) \cdot r(10^2) + r(60) && \pmod{11} \\ &= 6 \cdot 1 + 10 \cdot 1 + 5 && \pmod{11} \\ &= 6 + 10 + 5 && \pmod{11} \\ &= 10 && \pmod{11}. \end{aligned}$$

"By the way, Gnam, when computing 6 + 10 + 5 (mod 11), I mentally added the 6 and 5 first, obtaining 0, and then added the 10. That is, I thought

$$(6 + 10) + 5 = (6 + 5) + 10 = 0 + 10 = 10 \pmod{11}."$$

"That doesn't strike me as particularly shorter.

"Well, then, consider the following."

$$\begin{aligned} 7202163754748 &= 7(10^{12}) + 20(10^{10}) + 21(10^8) + 63(10^6) + 75(10^4) \\ &\quad + 47(10^2) + 48, \text{ so} \\ r(7202163754748) &= r(7) + r(20) + r(21) + r(63) + r(75) \\ &\quad + r(47) + r(48) \pmod{11} \\ &= 7 + 9 + 10 + 8 + 9 + 3 + 4 \pmod{11}. \end{aligned}$$

"Now, if you combine the 7 + 4 and the 8 + 3 you get 0's, so the above reduces to 9 + 10 + 9 (mod 11) or 6."

Exercise 2.12

Compute $r_{11}(38429)$, $r_{11}(1563219)$, and $r_{11}(896341285)$.

"I suppose that you also had a short way of obtaining the 6 from the 9 + 10 + 9."

"I did. I thought '9 = –2 (mod 11) and 10 = –1 (mod 11) and so 9 + 10 + 9 = –2 + –1 + –2 = –5 = 6 (mod 11)' "

"Where do the negative numbers come from?"

"Before answering that, let me back up a little. A few pages ago I introduced a system $\mathbf{Z}_{11}$ under the operations $+_{11}$ and $\cdot_{11}$. I didn't ask you to investigate the algebra of $\mathbf{Z}_{11}$ then because,

---

Answers, 2.12

I find it easier to do these if I 'read' numbers from right to left.

$r(38429) = r(29) + r(84) + r(3) = 7 + 7 + 3 = 6 \pmod{11}$;

$r(1563219) = r(19) + r(32) + r(56) + r(1) = 8 + 10 + 1 + 1 = 9 \pmod{11}$;

$r(896341285) = r(85) + r(12) + r(34) + r(96) + r(8) = 8 + 1 + 1 + 8 + 8 = 4 \pmod{11}$.

well, I didn't know whether you'd see any point to it. But when I was first introduced to the modular systems, I did investigate their algebraic properties out of curiosity. Because of these investigations, when I later came across problems which the modular systems could help me solve, I was able to apply the knowledge I had accumulated earlier.

"For example, I observed that for each a, b, and c in $\mathbf{Z}_{11}$, the following properties held.

Commutative Property of Addition:	$a + b = b + a$	(mod 11)
Associative Property of Addition:	$(a + b) + c = a + (b + c)$	(mod 11)
and the rather obvious Identity Property of Zero:	$0 + a = a$ and $a + 0 = a$	(mod 11).

"The Commutative and Associative Properties may also seem obvious to you, but remember, this isn't ordinary addition. It is the commutative and associative properties that allow me to regroup

$$7 + 9 + 10 + 8 + 9 + 3 + 4 \text{ to}$$

$$(7 + 4) + (8 + 3) + 9 + 10 + 9$$

and the identity property of 0 which motivates me to do so."

"If that was supposed to be an explanation of where the negatives came from, I must have missed something Ant."

"The point is that I paired numbers whose sum is 0. When the sum of two numbers is 0, the numbers are called **additive inverses.** 7 is the mod 11 additive inverse of 4, and 4 is the mod 11 additive inverse of 7. Similarly, 8 and 3 are mod 11 additive inverses. Every element b of $\mathbf{Z}_{11}$ has an additive inverse, namely $11 - b$.

"The statement '8 is the mod 11 additive inverse of 3' requires 37 characters. No self-respecting mathematician condones such extravagance. So we abbreviate the statement to '8 = –3 (mod 11)' which cuts the number of characters to 15."

Definition

> If $b + c = 0 \pmod{11}$ then b is called the **mod 11 additive inverse** of c. (Similarly, c is the mod 11 additive inverse of b.) **$-b$ (mod 11)** is an abbreviation for "the mod 11 additive inverse of b." You can also read $-b$ (mod 11) as "the additive inverse of b mod 11".

"Gnam, I want to emphasize that in the context of $\mathbf{Z}_{11}$, $-b$ does *not* refer to the integer $-b$, but rather to the *mod 11 additive inverse of* b. Each element of $\mathbf{Z}_{11}$ now has another name. –3 and 8 both name the same number, as do –5 and 6, and –0 and 0.

$$-8 = 3 \text{ and } -3 = 8 \text{ since } 3 + 8 = 0 \pmod{11};$$

$$-7 = 4 \text{ and } -4 = 7 \text{ since } 7 + 4 = 0 \pmod{11}.$$

"If saying 'the additive inverse of 5' is too much of a mouthful, you can say 'the opposite of 5' or just 'opp 5'. In the context of $\mathbf{Z}_{11}$, one does *not* read '–5' as 'negative 5', unless you're as disregardful of accuracy as a friend of mine, Andrea Rothbart, who not only stoops to saying 'negative 5', but regularly uses the outrageous expression 'minus 5' (which means 'subtract 5'). So try not to follow her poor example or allow yourself to be confused by it, if you should happen to meet her."

"I think you're being a bit too hard on her, Ant. After all, 'minus 5' does convey the notion that you're moving *backwards* 5 units on the clock from 0 and thus ending up at 6. I kind of like the sound of 'minus 5'. It has fewer syllables than most of the names you espouse. Yep, Rothbart sounds like my sort of mathematician. Introduce her to me sometime, will ya?"

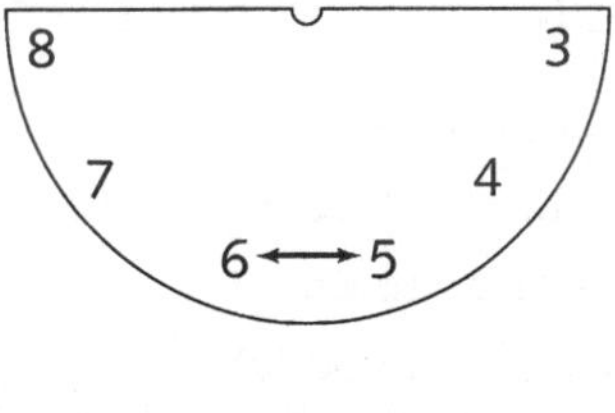

On an 11-hour clock, 5 is opposite 6 and 6 is opposite 5.

$5 = -6$ and $6 = -5$.

"As I was starting to explain, Gnam, using the mod 11 commutative and associative properties and the identity property of 0, along with the fact that $a + -a = 0$, one can show that for each a and b in $\mathbf{Z}_{11}$,

$$-a+(-b)=-(a+b) \quad (\text{mod } 11).$$

Thus,

$$\begin{aligned} 9+10+9 &= -2+(-1)+(-2) && (\text{mod } 11) \\ &= -(2+1+2) && (\text{mod } 11) \\ &= -5 && (\text{mod } 11) \\ &= 6 && (\text{mod } 11). \end{aligned}$$

"Since adding 0's, 1's and 2's is a lot easier than adding 9's and 10's, one of the strategies for doing modular arithmetic cleverly and quickly is to use names of numbers which are 'closer to 0'.

"By using additive inverses, we can further simplify the process of finding remainders upon division by 11. Note that $-1 \cdot 1 = 10 \cdot 1 = 10 = -1$ (mod 11) and $-1 \cdot 2 = 10 \cdot 2 = 9 = -2$ (mod 11). In general, $-1 \cdot b = -b$ (mod 11). (If you're skeptical, you can check out the remaining 9 instances).

"Also, $-1 \cdot (-1) = 10 \cdot 10 = 1$ (mod 11).

Now,

$$\begin{aligned} r(1) &= 1 && (\text{mod } 11) \\ r(10) &= 10 = -1 && (\text{mod } 11) \\ r(10^2) &= r(10)\cdot r(10) = -1\cdot(-1) = 1 && (\text{mod } 11) \\ r(10^3) &= r(10^2)\cdot r(10) = 1\cdot(-1) = -1 && (\text{mod } 11). \end{aligned}$$

In general, $r(10^k)= 1$ (mod 11) if k is even and $r(10^k) = -1$ (mod 11) if k is odd.

"Thus, to find $r_{11}(982)$ we can reason as follows:

$$982 = 9\cdot 100 + 8\cdot 10 + 2$$

so

$$\begin{aligned} r(982) &= 9\cdot 1 + 8\cdot(-1) + 2 && (\text{mod } 11) \\ &= 9 + (-8) + 2 && (\text{mod } 11) \\ &= 3 && (\text{mod } 11). \end{aligned}$$

Also,
$$8635 = 8 \cdot 10^3 + 6 \cdot 10^2 + 3 \cdot 10 + 5$$
so
$$\begin{aligned} r(8635) &= 8 \cdot (-1) + 6 \cdot 1 + 3 \cdot (-1) + 5 \pmod{11} \\ &= -8 + 6 + (-3) + 5 \pmod{11} \\ &= 0 \pmod{11}. \end{aligned}$$

"In general, to compute the remainder of b upon division by 11, just add mod 11 the digits of b, alternating the 'signs.' I find it easier to start with the ones digit:"

$$\begin{aligned} r(123456789) &= 9 + (-8) + 7 + (-6) + 5 + (-4) + 3 + (-2) + 1 \pmod{11} \\ &= (9+1) + [-8 + (-2)] + (7+3) + [-6 + (-4)] + 5 \pmod{11} \\ &= -1 + 1 + (-1) + 1 + 5 \pmod{11} \\ &= 5 \pmod{11}. \end{aligned}$$

Exercise 2.13

Without using the long division process, find the mod 11 remainder of each of the following: 348; –348; 60125; –60125; 420378591; –420378591; $123 \cdot 5918 - 78014 \cdot 152319$; $9183 \cdot 311254$.

Exercise 2.14

Find the mod 11 remainder of each of the following: 2442, 693396, and 81233218.

Exercise 2.15

For each of the following numbers, if possible fill in the blank with a single digit so that the resulting number is divisible by 11.

1. 2_4153
2. 785342_
3. 124365_69
4. 39_9750
5. 2_514

Answers, 2.13
7; 4; 10; 1; 6; 5; 7; 4

Answers, 2.14
$r(2442) = -2 + 4 + (-4) + 2 = [2 + (-2)] + [4 + (-4)] = 0 \pmod{11}$, $r(693396) = -6 + 9 + (-3) + 3 + (-9) + 6 = [6 + (-6)] + [9 + (-9)] + [3 + (-3)] = 0$ (mod 11), and $r(81233218) = -8 + 1 + (-2) + 3 + (-3) + 2 + (-1) + 8 = [8 + (-8)] + [1 + (-1)] + [2 + (-2)] + [3 + (-3)] = 0 \pmod{11}$.

These numbers are all palindromes with an even number of digits and so have a mod 11 remainder of 0.

Answers, 2.15
1. 7
2. 8
3. 7
4. 2
5. not possible

"You may have noticed that the algebraic structure of $\mathbf{Z}_{11}$ under $+_{11}$ and $\cdot_{11}$ bears a striking resemblance to the algebraic structure of the integers. Both are examples of what mathematicians call a **commutative ring.**[*] That is, both satisfy the following properties. For each a, b, and c in the set:

$a + b = b + a$	(Commutative Property of Addition)
$ab = ba$	(Commutative Property of Multiplication)
$(a + b) + c = a + (b + c)$	(Associative Property of Addition)
$(ab)c = a(bc)$	(Associative Property of Multiplication)
$a + 0 = 0 + a = a$	(Identity Property of zero)
$a \cdot 1 = 1a = a$	(Identity Property of one)
$a(b + c) = ab + \mathrm{ac}$	(The Distributive Property of Multiplication over Addition)
For each b in the set, there exists c in the set such that $c + b = b + c = 0$	(Existence of Additive Inverses).

"The above operations are mod 11 operations when we're referring to $\mathbf{Z}_{11}$. These properties can be used to simplify calculations. For example:

$$8 \cdot 6 = 4 \cdot (2 \cdot 6) = 4 \cdot 1 = 4 \pmod{11}$$
$$\text{and } 5 \cdot 8 + 6 \cdot 8 = (5 + 6) \cdot 8 = 0 \cdot 8 = 0 \pmod{11}.$$

"Indeed, one difference between man and robot is that man seeks clever ways of solving problems, whereas a robot typically fixes on one method and pursues it no matter how tedious the calculations are."

[*] Such a structure is also commonly called a commutative ring with identity (or unity). Both "commutative" and "identity" here refer to multiplication.

Exercise 2.16

All of the operations in this exercise are mod 11 operations. Use the number properties to find shortcuts. You also may find it helpful to substitute, for example, –1 for 10 and –2 for 9.

1. $4 \cdot 6 \cdot (-3) \cdot 2 \cdot 10$ (Hint: Look for products of 1 and –1.)
2. $10 \cdot 10 + 10 \cdot 9$
3. $4 \cdot 9 + 9 \cdot 6 + 3 \cdot 9$ (Hint: Use the distributive property.)
4. $8 \cdot 6 + (-5) \cdot 2 \cdot 6 \cdot 2$ (Hint: Think of 8 as $4 \cdot 2$ and look for products which are 1 or –1.)
5. $8 \cdot 8 + 7 \cdot 7 + 7 \cdot 8$
6. $8 + 5 + 7 + 6 + 4 + 9 + 10$
7. $8 \cdot 7 + (-9) \cdot 7 + 3 \cdot 7$
8. $9 \cdot 9 + (-8) \cdot (-8)$
9. $-7 + (-4) \cdot (-8) + (-10)$
10. $1 + 2 + 3 + 4 + 5 + 6 + 7 + 8 + 9 + 10$
11. $-10 \cdot (-9) \cdot (-8)$

"The substitution of the minuses really helped, Ant! Instead of multiplying 9 by 8, I multiplied –2 times –3 and got the answer 6 much more easily."

"In any commutative ring, properties such as $(-a)(-b) = ab$ and $(-a)b = a(-b) = -(ab)$ hold, Gnam."

Supplementary Problem 2.3

Suppose a, b, c, and k are elements of a commutative ring. Prove the following.

1. If $k + x = x + k = x$ for each element x of the ring, then $k = 0$. (Hint: What sum do you get when you add $0 + k$?)
2. If for some element x, there are elements d and d' such that $x + d = d + x = 0$ and $x + d' = d' + x = 0$, then $d = d'$. (Hint: What is $d + (x + d')$? What is $(d + x) + d'$?)

Answers, 2.16

1. $4 \cdot -3 \cdot 6 \cdot 2 \cdot 10 = -1 \cdot 1 \cdot -1 = 1$
2. $1 + -9 = -8 = 3$
3. $(4 + 6 + 3)9 = 2 \cdot -2 = -4 = 7$
4. $4 + 1 = 5$
5. 4
6. 5
7. 3
8. 2
9. 4
10. 0
11. $1 \cdot 2 \cdot 3 = 6$

3. If $k \cdot x = x \cdot k = x$ for each element x of the ring then $k = 1$. (Hint: Use a strategy similar to the one you used to prove part 1.)
4. $a \cdot 0 = 0$ [Hint: $a \cdot 0 = a(0 + 0)$.]
5. $-1a = -a$ [Hint: To prove that stuff = –(junk), it suffices to show that stuff + junk = 0.]
6. $-a + (-b) = -(a + b)$.
7. $(-a)b = a(-b) = -(ab)$.
8. $-(-a) = a$.
9. $(-a)(-b) = ab$.
10. If $a + b = a + c$ then $b = c$.

Supplementary Problem 2.4

Prove that $\mathbf{Z}_{11}$ under mod 11 addition and multiplication is a commutative ring. (Suggestion: Use the definitions of mod 11 addition and multiplication, along with the fact that the integers form a commutative ring.)

Exercise 2.17

All of the operations in this exercise are mod 11 operations. Fill in the blanks to make the sentences true. Try to avoid unnecessary calculations by making use of ring properties and shortcuts.

1. $\_\_ \cdot 10^7 = 3$.
2. $4 \cdot 7 + \_\_ \cdot 7 = 2 \cdot 7$.
3. $10 \cdot 7 + 1 \cdot 5 = \_\_ \cdot 10$.
4. $6 \cdot 4 + (-5) \cdot 3 = \_\_ \cdot 5$.
5. $2 \cdot 8 + 8 \cdot \_\_ = 1 \cdot 8$.
6. $6 \cdot 5 \cdot \_\_ \cdot 2 = 1$.
7. $5 \cdot 9 + (-8) \cdot 9 = 9 \cdot \_\_\_$.

"Let's move away from arithmetic into big time stuff: algebra. As you will see, Gnam, the modular systems are highly powerful tools for solving a variety of problems. Finding solutions commonly requires setting up and then solving equations. Since $\mathbf{Z}_{11}$ is finite, one way to solve equations is to substitute each element of $\mathbf{Z}_{11}$ into the equation to see if it is a solution. But a more elegant way of solving equations is to make use of the algebraic structure of $\mathbf{Z}_{11}$."

Answers, 2.17

1. 8
2. 9
3. 2
4. 4 (think of $6 \cdot 4$ as $-5 \cdot 4$)
5. 10
6. 9
7. 8

Notation

"⇔" is an abbreviation for "if and only if".

$$
\begin{array}{lrl}
 & 3x+9=3 & \pmod{11} \\
\Leftrightarrow & (3x+9)+2=3+2 & \pmod{11} \\
\Leftrightarrow & 3x=5 & \pmod{11} \\
\Leftrightarrow & 4(3x)=4(5) & \pmod{11} \\
\Leftrightarrow & x=9 & \pmod{11}.
\end{array}
$$

"Hence, 9 is the unique solution to the equation $3x + 9 = 3$ (mod 11). Note that our first step was to add 2 to both sides of the equation to isolate $3x$ (in ordinary algebra we would subtract 9). We then multiplied both sides by 4 to isolate x (in ordinary algebra we would multiply by $1/3$)."

Exercise 2.18

Solve each of the following equations.

1. $x+9=4$ (mod 11).
2. $6x=7$ (mod 11).
3. $3x+7=4$ (mod 11).
4. $2=5x+(-4)$ (mod 11).
5. $2x+5x+8=1+4$ (mod 11).
6. $8x=8x+1$ (mod 11).
7. $2x=-9x$ (mod 11).

"The hardest part of solving these equations, Ant, is finding the number to multiply both sides of the equation by to get x by itself."

"To isolate x, you need to multiply ax by some number to obtain $1x = x$. When the product of two numbers is 1, the numbers are called **multiplicative inverses**. For example, 2 and 6 are mod 11 multiplicative inverses since $2 \cdot 6 = 1$ (mod 11). Similarly, 3 is the mod 11 multiplicative inverse of 4 and 4 is the mod 11 multiplicative inverse of 3."

Answers, 2.18

1. $x+9=4 \Leftrightarrow x+9+2=4+2 \Leftrightarrow x=6$ (mod 11).
2. $6x=7 \Leftrightarrow 2(6x)=2{\cdot}7 \Leftrightarrow x=3$ (mod 11).
3. $3x+7=4 \Leftrightarrow 3x=-3 \Leftrightarrow x=-1 \Leftrightarrow x=10$ (mod 11).
4. $2=5x+-4 \Leftrightarrow 6=5x \Leftrightarrow 9(6)=9(5x) \Leftrightarrow x=10$ (mod 11).
5. $7x+8=5 \Leftrightarrow 7x=-3 \Leftrightarrow 8(7x)=8(-3) \Leftrightarrow x=9$ (mod 11).
6. $8x=8x+1$ has no solutions.
7. Since -9 is another name for 2, this is the equation $2x=2x$ (mod 11) and every element of $\mathbf{Z}_{11}$ is a solution!

Definition

> If $bc = 1 \pmod{11}$ then b is called the **mod 11 multiplicative inverse of c.** (Similarly, c is the mod 11 multiplicative inverse of b.) **'/b (mod 11)'** is an abbreviation for 'the mod 11 multiplicative inverse of b'. You can read '$/b$' as 'the multiplicative inverse of b' or as '**slash b**'.
>
> Examples: 7 and 8 are mod 11 multiplicative inverses since $7 \cdot 8 = 1 \pmod{11}$. Hence, $/7 = 8 \pmod{11}$. Similarly $/8 = 7 \pmod{11}$. Also, $/6 = 2$ and $/2 = 6 \pmod{11}$.

"Do you see the analogy between 0 and additive inverses on one hand, and 1 and multiplicative inverses on the other? 0 is my very favorite number to add, since $0 + a = a$ for any a. For being so accommodating (under addition), 0 is accorded a special title: it is called the **additive identity**. When adding, we pray for as many 0's as possible. In particular, we seek pairs of numbers whose sum is 0. Such numbers are called **additive inverses.**

"Well, 1 plays an analogous role under multiplication, since $1a = a$ for any a. Hence, 1 is called the **multiplicative identity**, and pairs of numbers whose product is 1 are called **multiplicative inverses**. Now, 0, who is so passive under addition, becomes a lion under multiplication since $0a = 0$ for every a. Thus, 0 has no mod 11 multiplicative inverse. But each of the other elements of $\mathbf{Z}_{11}$ does."

Exercise 2.19

> Find the mod 11 multiplicative inverse of each nonzero element of $\mathbf{Z}_{11}$.

"In $\mathbf{Z}_{11}$ another name for 6 is $/2$. So, whenever you see a 6 you can substitute $/2$ (or -5) if it suits your fancy. As we solve equations over $\mathbf{Z}_{11}$, it may be useful to refer to a list of mod 11 multiplicative inverses.

"Consider the equation, $2x - 3 = 4 \pmod{11}$. It involves the operation *subtraction*. How should we define subtraction over

---

Answers, 2.19

$/1 = 1$;	$/2 = 6$;	$/3 = 4$;	$/4 = 3$;	$/5 = 9$;
$/6 = 2$;	$/7 = 8$	$/8 = 7$;	$/9 = 5$;	$/10 = 10$.

$\mathbf{Z}_{11}$? There are several equivalent ways*, but here is the one I prefer."

Definition

If a and b are elements of $\mathbf{Z}_{11}$, $a - b$ means $a + (-b)$ (mod 11).

Examples: $8 - 3 = 5$ (mod 11) since $8 - 3 = 8 + (-3) = 8 + 8 = 5$ (mod 11). (Moving backward 3 hours on an 11-hour clock is equivalent to moving forward 8 hours.) Similarly, $6 - 10 = 6 + (-10) = 6 + 1 = 7$ (mod 11). (Moving backward 10 is equivalent to moving forward 1.)

"Even better, $6 - 10$ (in any reasonable world) is -4. And $\mathbf{Z}_{11}$ is eminently reasonable. So you can think: $6 - 10 = -4 = 7$ (mod 11)."

Exercise 2.20

Compute mod 11:

1. $5 - 3$ **2.** $6 - 8$ **3.** $0 - 4$ **4.** $5 - 10$

Back to equations:

$$
\begin{aligned}
& & 5-2(4-x) &= 8(x+7)+3x && \text{(mod 11)}\\
&\Leftrightarrow & 5-8+2x &= 8x+1+3x && \text{(mod 11)}\\
&\Leftrightarrow & -3+2x &= 1 && \text{(mod 11)}\\
&\Leftrightarrow & 2x &= 4 && \text{(mod 11)}\\
&\Leftrightarrow & x &= 2 && \text{(mod 11).}
\end{aligned}
$$

"In general, you can pretty much treat equations over $\mathbf{Z}_{11}$ as you treat them in ordinary algebra. $\mathbf{Z}_{11}$ under $+_{11}$ and $\cdot_{11}$ has all those nice properties you've come to rely upon, so when solving equations, just go ahead and do what seems to come naturally (at least as long as what comes naturally works!)" **

Answers, 2.20

1. 2 2. −2, or 9 3. −4, or 7 4. −5, or 6

* Can you think of one or two other equivalent ways to define subtraction?

** One notion that doesn't extend to $\mathbf{Z}_{11}$ is that of inequality. There is no way to define "less than" on the elements of $\mathbf{Z}_{11}$ which satisfies: If $a < b$ then $a+c < b+c$. You might want to think about why.

Another example:

$$
\begin{aligned}
& 4x-3(x-2)=x+5 \pmod{11}\\
\Leftrightarrow\quad & 4x-3x+6=x+5 \pmod{11}\\
\Leftrightarrow\quad & 6=5 \pmod{11}.
\end{aligned}
$$

But 6 does not equal 5, so the equation has no solutions.

Exercise 2.21

Solve each of the following equations.

1. $5x-7=8 \pmod{11}$.
2. $3(x+2)-5(x-1)=2(x+3) \pmod{11}$.
3. $6-7(x+5)=10 \pmod{11}$.
4. $7(x+1)+5(x+2)=x-1 \pmod{11}$.
5. $3(x+2)-4(5-x)=8-4x \pmod{11}$.
6. $7(2x-1)+3x=5(2-4x) \pmod{11}$.

"Another common problem in ordinary algebra is to solve two or more linear equations simultaneously. We can use the same techniques to solve systems of equations over $\mathbf{Z}_{11}$. There are several such techniques. One is illustrated below."

1. $3x+5y=10 \pmod{11}$.

2. $2x+3y=4 \pmod{11}$.

Multiplying equation 1 by 8, we obtain the following equivalent system.

1′. $2x+7y=3 \pmod{11}$.

2. $2x+3y=4 \pmod{11}$.

Subtracting equation 2 from 1′, we obtain

$$4y=10 \pmod{11}.$$

Hence $$y=8 \pmod{11}.$$

Substituting 8 for y in either one of the first two equations, we obtain $x=1$. So the pair $(x,y)=(1,8)$ is the solution of the original system of equations.

Answers, 2.21

1. 3	2. 4	3. 7
4. no solutions	5. All elements of $\mathbf{Z}_{11}$ are solutions.	6. 7

Exercise 2.22

Find all pairs of numbers x and y in $\mathbf{Z}_{11}$ (if any exist) that satisfy the following systems of equations.

1. $5x + y = 10$ and $3x - y = 5 \pmod{11}$.
2. $6x + 2y = 1$ and $4x + 9y = 4 \pmod{11}$. (Suggestion: Just add these two equations!)
3. $4x + 5y = 3$ and $5x + 2y = 7 \pmod{11}$.
4. $2x - 7y = 4$ and $3x + 4y = 4 \pmod{11}$.
5. $6x - 2y = 2$ and $x + 3y = 6 \pmod{11}$.
6. $4x - 5y = 6$ and $5x + 2y = 2 \pmod{11}$.
7. $3x + 5y = 2$ and $5x + y = 1 \pmod{11}$.

Supplementary Problem 2.5

The graph of an equation of the form $ax + by = c$ over the real number system is a line (unless a and b are both 0). Graph the set of points described by, say, the equation $2x + 3y = 7 \pmod{11}$. What does the graph look like? Might it make sense to extend the concept of slope to this graph? If so, what would be its slope(s)?

"You've shown us how to add, subtract and multiply in $\mathbf{Z}_{11}$, Ant. But you have overlooked one of the big four operations. You began this manuscript with a tirade about division, and now you completely ignore the operation!"

"In a reasonable algebraic structure such as the real number system or $\mathbf{Z}_{11}$, division, like subtraction, is a redundant operation. In any commutative ring, subtraction really isn't 'new': $a - b$ is just $a + (-b)$. Similarly, if an element b of the commutative ring has a multiplicative inverse, division isn't 'new': $a \div b$ is just $a \cdot (/b)$, or more concisely, a/b. So, for example

$$2 \div 3 = 2/3 = 2 \cdot 4 = 8 \pmod{11}.$$

Answers, 2.22

1. (6, 2) 2. (6, 10) 3. (3, 7)
4. (0, 1) 5. (2, 5)
6. There are 11 solutions: (0, 1), (1, 4), (2, 7), (3, 10), (4, 2), *etc.* In general, any pair of the form $(k, 1 + 3k)$ is a solution where k is an element of $\mathbf{Z}_{11}$ [that is, $1 + 3k \pmod{11}$, of course].
7. no solutions

However, because the division sign ÷ is so clumsy to work with, I'm not going to use it. But if you like, rather than reading 'a/b' as 'a times slash b' (or 'a times the multiplicative inverse of b'), you can read it as 'a divided by b'. Personally, I prefer to say 'a slash b'. (Fewer syllables.)"

Exercise 2.23

The following questions pertain to $\mathbf{Z}_{11}$.

1. Show that $1/2 + 1/2 = 1$.
2. Show that $1/3 + 1/2 = 5/6$.
3. Compute $5/4 + 8/3$.
4. Compute $6/7 - 3/2$.
5. Is $6/9 = 2/3$? Is $2/4 = 1/2$?
6. Show that $(5/2)(2/3) = 5/3$.
7. Compute $(6/3)/(2/9)$.
8. Compute $(7/4)/(3/2) + (4/5)$.
9. Compute $(2/3)(3/2)$.

Exercise 2.24

Solve each of the following equations over $\mathbf{Z}_{11}$.

1. $x/3 + x/2 = 5$.
2. $2x/4 - 7/2 = x/3$.
3. $(x + 1)/3 = (5x - 2)/7$.
4. $5/(3x) = /(4 + x)$.
5. $1 + 3/x = -8/x$.
6. $(4 + x)/(x + 3) = 5$.

Answers, 2.23

1. $1/2 + 1/2 = 6 + 6 = 1$
2. $1/3 + 1/2 = 1 \cdot 4 + 1 \cdot 6 = 10 = 5 \cdot 2 = 5/6$
3. 3
4. 8
5. Yes, 8; Yes, 6
6. $(5/2)(2/3) = 5 \cdot /2 \cdot 2 \cdot /3 = 5/3$
7. 9
8. 6
9. 1

Answers, 2.24

1. $4x + 6x = 5 \Leftrightarrow -x = 5 \Leftrightarrow x = -5 \Leftrightarrow x = 6 \pmod{11}$.
2. $x/2 - 7/2 = 4x \Leftrightarrow x - 7 = 8x \Leftrightarrow 7x = -7 \Leftrightarrow x = -1 \Leftrightarrow x = 10 \pmod{11}$.
3. $7x + 7 = 4x - 6 \Leftrightarrow 3x = -2 \Leftrightarrow x = -8 \Leftrightarrow x = 3 \pmod{11}$.
4. $x = 1$.
5. This equation has no solutions. Recall that 0 has no multiplicative inverse.
6. $x = 0$.

Supplementary Problem 2.6

Prove that the following properties hold for $\mathbf{Z}_{11}$ under $+_{11}$ and $\cdot_{11}$.

1. $(a/b)(b/a) = 1$ when $a \neq 0$ and $b \neq 0$.
2. $/(a/b) = b/a$ when $a \neq 0$ and $b \neq 0$.
3. $(/a)(/b) = /(ab)$ when $a \neq 0$ and $b \neq 0$. [Suggestion: stuff $=$ /junk $\Leftrightarrow$ stuff $\cdot$ junk $= 1$.]
4. $(a/b)(c/d) = (ac)/(bd)$ when $b \neq 0$ and $d \neq 0$.
5. $a/b + c/d = (ad + bc)/(bd)$ when $b \neq 0$ and $d \neq 0$.
6. $a/b = c/d \Leftrightarrow ad = bc$ when $b \neq 0$ and $d \neq 0$.
7. $(a/b)/(c/d) = (ad)/(bc)$ when $b \neq 0$, $c \neq 0$, and $d \neq 0$.
8. $(ac)/(bc) = a/b$ when $b \neq 0$ and $c \neq 0$.

"I think that we've pretty much covered ninth-grade algebra, Ant."

"Not quite, Gnam. For example, we haven't solved any quadratic equations. Recall that in the real number system, the real solutions of the equation $ax^2 + bx + c = 0$, $a \neq 0$, are

$$\frac{-b \pm \sqrt{b^2 - 4ac}}{2a}$$

provided that $b^2 - 4ac \geq 0$. This formula is called the **quadratic formula**.

"The common method of deriving this formula for quadratic equations over the reals also works for quadratic equations over $\mathbf{Z}_{11}$ until one reaches the point where square roots must be taken. In $\mathbf{Z}_{11}$, it isn't clear what we would mean by, say, $\sqrt{4}$."

"Yes it is. $\sqrt{4} = 2$. It just isn't clear what you'd mean by $\sqrt{3}$, Ant."

"But there are *two* elements in $\mathbf{Z}_{11}$ whose squares are 4, Gnam: 2 and 9, since $2^2 = 9^2 = 4 \pmod{11}$. So which square root would we mean by $\sqrt{4}$? 2 or 9? Similarly, there are two elements of $\mathbf{Z}_{11}$ whose square is 3: $5^2 = 6^2 = 3 \pmod{11}$. So, which square root would $\sqrt{3}$ refer to? Rather than assigning meaning to $\sqrt{k} \pmod{11}$, let's just restate the quadratic formula so that it applies to $\mathbf{Z}_{11}$."

Quadratic formula

> Suppose that $a \neq 0$. Then x_0 is a solution of the equation $ax^2 + bx + c = 0 \pmod{11}$ if and only if $x_0 = (-b + k)/(2a)$ where k is a square root of $b^2 - 4ac \pmod{11}$.

"For example, to find the solutions of $x^2 + 3x + 1 = 0 \pmod{11}$, we could reason as follows: $b^2 - 4ac = 3^2 - 4 \cdot 1 \cdot 1 = 9 - 4 = 5 \pmod{11}$. Now, 4 and 7 are the (only) mod 11 square roots of 5 [$4^2 = 5$ and $7^2 = 5 \pmod{11}$]. By substituting 4 and 7 for k in the formula, we obtain the solutions of the equation $x^2 + 3x + 1 = 0 \pmod{11}$:"

$$(k = 4) \quad x_0 = (-3 + 4)/(2 \cdot 1) = 1/2 = 6 \qquad \text{(mod 11) and}$$

$$(k = 7) \quad x_0 = (-3 + 7)/(2 \cdot 1) = 2 \qquad \text{(mod 11).}^*$$

"How did you know what the square roots of 5 were, Ant?"

"Let's make up a table of mod 11 squares."

a	0	1	2	3	4	5	6	7	8	9	10
a^2 (mod 11)	0	1	4	9	5	3	3	5	9	4	1

"Note that there are six perfect squares in $\mathbf{Z}_{11}$: 0, 1, 3, 4, 5, and 9, and that each of these but 0 has two square roots. According to the table, the square roots of 5 are 4 and 7. Note also that, ignoring 0, the bottom row of the table is the same reading to the right as reading to the left. This is because $b^2 = (-b)^2$. If b is a square root of some number, $-b$ must also be a square root of that number.

"As another example, consider $2x^2 + 4x + 1 = 0 \pmod{11}$. $b^2 - 4ac = 5 - 8 = 8 \pmod{11}$. 8 is not a perfect square, so the equation $2x^2 + 4x + 1 = 0$ has no solutions in $\mathbf{Z}_{11}$.

"Hence, we have the same situation as when we solve quadratic equations within the framework of the real number system. In ordinary algebra, when $b^2 - 4ac$ is negative, we throw up our arms and exclaim that the world of real numbers is too small to allow us to solve the equation. This occurs roughly half of the time. Similarly, when we operate within the framework of $\mathbf{Z}_{11}$, roughly half of the time quadratic equations will have no solutions.

* This means that $x^2 + 3x + 1$ can be factored into $(x - 6)(x - 2)$ over $\mathbf{Z}_{11}$.

"Consider the quadratic $x^2 + 6x + 8$. You may see that it can be factored into $(x + 4)(x + 2)$. Rather than using the quadratic formula to solve $x^2 + 6x + 8 = 0$, you could argue as follows."

$$x^2 + 6x + 8 = 0 \pmod{11}$$
$$\Leftrightarrow \quad (x + 4)(x + 2) = 0 \pmod{11}$$
$$\Leftrightarrow \quad x + 4 = 0 \text{ or } x + 2 = 0 \pmod{11}$$
$$\Leftrightarrow \quad x = 7 \text{ or } x = 9 \pmod{11}.$$

"I like that method better than using the quadratic formula, Ant. And you're right. I may see how to factor a quadratic. But it's also true, and far more likely, that I may not. I just read the previous footnote, and it would never occur to me to factor $x^2 + 3x + 1$ into $(x - 6)(x - 2)$. That's just plain silly!"

"Well then, when you want to solve a quadratic, forget the formula. Instead, 'complete the square' by adding the square of half the coefficient of x to both sides of the equation. For example, to solve $3x^2 - 6x - 1 = 0$, you might reason as follows."

$$3x^2 - 6x = 1 \pmod{11}$$
$$\Leftrightarrow \quad x^2 - 2x = 4 \pmod{11} \quad \text{[multiply by 4]}$$
$$\Leftrightarrow \quad x^2 - 2x + 1 = 4 + 1 \pmod{11} \quad \text{[add } (-2/2)^2 = 1 \text{ to both sides]}$$
$$\Leftrightarrow \quad (x - 1)^2 = 5 \pmod{11}.$$

"Now, the mod 11 square roots of 5 are 4 and 7. Hence,

$$x - 1 = 4 \quad \text{or} \quad x - 1 = 7 \pmod{11}.$$

So:

$$x = 5 \quad \text{or} \quad x = 8 \pmod{11}.$$

Check: $3(5)^2 - 6(5) - 1 = 3(3) - 8 - 1 = 0 \pmod{11}$

and $3(8)^2 - 6(8) - 1 = 3(9) - 4 - 1 = -6 - 4 - 1 = 0 \pmod{11}$.

"Here's another example."

$$6x^2 - 7x + 5 = 0 \pmod{11}$$
$$\Leftrightarrow \quad x^2 - 3x = -10 \pmod{11}$$
$$\Leftrightarrow \quad x^2 + 8x = -10 \pmod{11} \quad \text{["I'd rather halve 8 than } -3.\text{"]}$$
$$\Leftrightarrow \quad x^2 + 8x + 5 = -10 + 5 \pmod{11} \quad \text{[add } 4^2 = 5 \text{ to both sides]}$$
$$\Leftrightarrow \quad (x + 4)^2 = 6 \pmod{11}.$$

"Now, 6 is not a perfect square in $\mathbf{Z}_{11}$, so $6x^2 - 7x + 5 = 0$ has no solutions."

Exercise 2.25

Using any method you like, solve each of the following equations over $\mathbf{Z}_{11}$.

1. $x^2 - 7x + 3 = 0.$
2. $x^2 + 1 = 0.$
3. $x^2 + 7 = 0.$
4. $x^2 + 5x + 6 = 0.$
5. $3x^2 - 6x + 2 = 0.$
6. $3x^2 + 4x = 0.$
7. $5x^2 - 7 = 0.$
8. $4x(5 - x) = 2(x - 3).$
9. $4 - x^2 = 3x.$
10. $x^2 - 5x = 5.$
11. $(2x + 1)(3x - 1)(7x + 2) = 0.$
12. $(x + 1)(x - 9) = 2x(x - 3).$
13. $(3/7)x^2 + (4/5)x - 2 = 0.$
14. $(3x - 2)(5x - 7) = (2x + 3)(5 - 9x).$

Supplementary Problem 2.7

Factor the quadratics in parts 1, 3, 4, and 5 above into the product of two linear polynomials over $\mathbf{Z}_{11}$.

"So much for $\mathbf{Z}_{11}$ for now. I still have infinitely more algebraic systems that I want to present."

"Then I suggest you discontinue your practice of introducing them one at a time."

"Very well, I'll introduce the rest of them all at once. I'll more or less breeze through the presentation since the definitions and notation are merely extensions of the ones we used for $\mathbf{Z}_{11}$. Once you understand $\mathbf{Z}_{11}$, an understanding of the other modular systems comes with little extra effort, except, well there is a surprise or two in store. Not all of the modular systems are as reasonable as $\mathbf{Z}_{11}$."

Answers, 2.25

1. 8 and 10 — 2. no solutions — 3. 2 and 9
4. 8 and 9 — 5. 3 and 10 — 6. 6 and 0
7. no solutions — 8. no solutions — 9. 1 and 7
10. 2 and 3
11. When the product of three numbers is 0 in $\mathbf{Z}_{11}$, one of the factors must be 0. Hence the solutions are 4, 5, and 6.
12. 4 and 5 — 13. 6 and 9
14. This turns out to be a linear equation with solution 7.

Note: *In the following,* a, b, *and* n *refer to integers.*

Notation

If $n > 1$, $\boldsymbol{r_n(a)}$ means the remainder when a is divided by n.

Examples: $r_5(12) = 2$; $r_7(73) = 3$; $r_3(-8) = 1$.

Exercise 2.26

Find the following.

1. $r_{15}(-32)$	**2.** $r_9(41)$	**3.** $r_7(8)$
4. $r_{19}(193)$	**5.** $r_{12}(-11)$	**6.** $r_{10}(1078968)$
7. $r_2(-103)$	**8.** $r_{16}(32649)$	

Exercise 2.27

Suppose that n is a factor of a where $n > 1$. What can you conclude about $r_n(a)$? Suppose that $r_n(a) = r_n(b)$. What can you conclude about the relationship between a and b?

Notation

For each $n > 1$, $\mathbf{Z}_n$ denotes the set of remainders upon division by n. That is, $\mathbf{Z}_n = \{0, 1, 2, 3, \dots, n-1\}$.

Definition

For each $n > 1$, $+_n$ and $\cdot_n$ are operations on $\mathbf{Z}_n$ defined as follows:

$$a +_n b = r_n(a + b)$$
$$a \cdot_n b = r_n(ab).$$

Examples: $7 +_9 6 = r_9(13) = 4$; $5 \cdot_8 4 = r_8(20) = 4$;
$1 +_2 1 = 0$; $3 \cdot_5 2 = 1$.

For each $n > 1$, $\mathbf{Z}_n$ under the operations $+_n$ and $\cdot_n$ is called a **modular system.**

Answers, 2.26

1. 13	2. 5	3. 1	4. 3
5. 1	6. 8	7. 1	8. 9

Answers, 2.27

n is a factor of $a \Leftrightarrow r_n(a) = 0$; $r_n(a) = r_n(b) \Leftrightarrow n$ is a factor of $a - b$.

Exercise 2.28

Compute the following.

1. $8 +_{10} 9$ 2. $7 \cdot_8 7$ 3. $5 +_6 3$
4. $8 \cdot_{12} 11$ 5. $8 +_{15} 5 +_{15} 10 +_{15} 7$ 6. $10 \cdot_{20} 4$

*Super-Important Theorem**

If $n > 1$ and a and b are integers, then

$$r_n(a+b) = r_n(a) +_n r_n(b) \text{ and}$$
$$r_n(ab) = r_n(a) \cdot_n r_n(b).$$

Exercise 2.29

Use the Super-Important Theorem to compute each of the following.

1. $r_7(77 + 15)$ 2. $r_{15}(45 + 9)$
3. $r_3(999 \cdot 32)$ 4. $r_6(14 \cdot 10 + 3)$
5. $r_{91}(91 \cdot 15 \cdot 789)$ 6. $r_{10}(78 + 39)$
7. $r_6(-12 + [-2])$ 8. $r_{10}(-15 \cdot 37)$
9. $r_{15}(30 \cdot 84 + 46 \cdot 16)$ 10. $r_{16}(33 \cdot 19 - 8 \cdot 48 + 162)$
11. $r_{45}(92 \cdot 92 + 48 \cdot 2)$

Notation

We will omit the subscript n on $+_n$, $\cdot_n$, and r_n and instead write (mod n) to the right of expressions, equations, and sequences of equations when we wish the subscript n to be understood.

Examples: $r(12) + r(15) \pmod 6$ means $r_6(12) +_6 r_6(15)$ and
$4 + 6 = 3 \pmod 7$ means $4 +_7 6 = 3$.

Answers, 2.28

1. 7 2. 1 3. 2
4. 4 5. 0 6. 0

* A function which maps one commutative ring onto another in a way which "preserves" addition and multiplication is called a morphism. In particular, for each $n > 1$, r_n is a morphism from the set of integers, **Z**, onto $\mathbf{Z}_n$. Perhaps a better name for this theorem is "The Morphism Theorem."

Answers, 2.29

1. $0 +_7 1 = 1.$ 2. $0 +_{15} 9 = 9.$ 3. $0 \cdot_3 2 = 0.$
4. $2 \cdot_6 4 +_6 3 = 5.$ 5. $0 \cdot_{91}$ (who cares) $= 0.$ 6. $8 +_{10} 9 = 7.$
7. $0 +_6 4 = 4.$ 8. $5 \cdot_{10} 7 = 5.$ 9. $0 +_{15} 1 \cdot_{15} 1 = 1.$
10. $1 \cdot_{16} 3 -_{16} 0 +_{16} 2 = 5.$
11. $2 \cdot_{45} 2 +_{45} 3 \cdot_{45} 2 = 10.$

Each $\mathbf{Z}_n$ under $+_n$ and $\cdot_n$ is a commutative ring. That is:

For each a, b, and c in $\mathbf{Z}_n$

$a+b=b+a$	(Commutative Property of Addition)
$ab=ba$	(Commutative Property of Multiplication)
$(a+b)+c=a+(b+c)$	(Associative Property of Addition)
$(ab)c=a(bc)$	(Associative Property of Multiplication)
$a+0=0+a=a$	(Identity Property of zero)
$a\cdot 1=1a=a$	(Identity Property of one)
$a(b+c)=ab+ac$	(The Distributive Property of Multiplication over Addition)
For each b in the set, there exists c in the set such that $c+b=b+c=0$	(Existence of Additive Inverses).

"As usual, Gnam, **the mod n additive inverse of b is denoted by $-b$ (mod n).** So, for example, $-10=5$ (mod 15) but $-10=7$ (mod 17). Also, we define mod n subtraction in terms of addition, although, if you prefer, you can think of mod n subtraction as moving counterclockwise on an n-hour clock."

Definition

$a-b$ (mod n) means $a+(-b)$ (mod n).

"It seems to me that everything that happened in mod 11 happens with all the mods. So what are those mysterious differences you referred to, Ant?"

"They'll be uncovered in the course of our investigation. Let's begin with some clever arithmetic computations."

Example: Compute $23+30+9+18+27$ (mod 31).

$$
\begin{aligned}
23+30+9+18+27 &= -8+(-1)+9+(-13)+(-4) && \text{(mod 31)}\\
&= (-8+[-1]+9)+([-13]+[-4]) && \text{(mod 31)}\\
&= 0+(-13)+(-4) && \text{(mod 31)}\\
&= -17 && \text{(mod 31)}\\
&= 14 && \text{(mod 31).}
\end{aligned}
$$

"Note the strategy of expressing a number as the additive inverse of another number when it simplifies computation. Also, seek sums which are 0 or close to 0."

Exercise 2.30

Compute as cleverly as possible (or at least as cleverly as occurs to you).

1. $26 + 18 + 14 + 15 + 27 \pmod{28}$
2. $98 + 15 + 11 + 97 + 5 \pmod{101}$
3. $2 \cdot 60 + 59 \cdot 59 \pmod{62}$
4. $1 + 2 + 3 + \ldots + 18 + 19 + 20 \pmod{21}$
5. $20 \cdot 28 \pmod{30}$
6. $58 \cdot 59 \cdot 60 \pmod{61}$
7. $3091 \cdot 3093 \pmod{3095}$

"Below is another example of my wits at work, Gnam. To compute $24 \cdot 4 \pmod{31}$, I reason as follows."

$$24 \cdot 4 = 3 \cdot (8 \cdot 4) = 3 \cdot 1 = 3 \pmod{31}.$$

"What wits? What you did was no easier than just finding the mod 31 remainder of $24 \cdot 4$ or 96. And 96 mod 31 is 3 since 96 is 3 more than 93."

"Well, then, consider the following."

$$\text{Compute } 3 \cdot 10 \cdot 16 \cdot 18 \cdot 3 \pmod{31}.$$

$$\begin{aligned} 3 \cdot 10 \cdot 16 \cdot 18 \cdot 3 &= (3 \cdot 10) \cdot (16 \cdot 2) \cdot (9 \cdot 3) &\pmod{31} \\ &= -1 \cdot 1 \cdot 27 &\pmod{31} \\ &= -27 &\pmod{31} \\ &= 4 &\pmod{31}. \end{aligned}$$

"Observe the strategy I am using. I try to combine factors that make a product of 1 or –1, or failing that, products as close to 0 as possible."

Answers, 2.30

1. $-2 + 18 + 1 + -1 = 16 \pmod{28}$.
2. $-3 + 26 + -4 + 5 = -2 + 26 = 24 \pmod{101}$.
3. $2 \cdot -2 + -3 \cdot -3 = -4 + 9 = 5 \pmod{62}$.
4. $(1 + 20) + (2 + 19) + (3 + 18) + (\text{etc}) = 0 + 0 + 0 + \ldots + 0 = 0 \pmod{21}$.
5. $-10 \cdot -2 = 20 \pmod{30}$.
6. $-3 \cdot -2 \cdot -1 = -6 = 55 \pmod{61}$.
7. $-4 \cdot -2 = 8 \pmod{3095}$.

"You must be kidding, Ant! I could take out my trusty calculator, compute the product of that mess, then...."

"But then you would miss out on the great reward of using your human-like brain to cleverly find shortcuts to calculations."

"Personally, I prefer being human-like to being Ant-like. So supply me with a calculator problem, please."

Exercise 2.31

Figure out how to use a calculator to find the mod 31 product

$$3 \cdot 10 \cdot 16 \cdot 18 \cdot 3.$$

Exercise 2.32

Use your ingenuity, rather than a calculator, to compute the following.

1. $15 \cdot 12 \cdot 7 \cdot 29 \cdot 8 \pmod{31}$
2. $11 \cdot 17 \cdot 26 \cdot 15 \pmod{33}$
3. $17 \cdot 12 + 12 \cdot 12 \pmod{30}$
4. $16 \cdot 15 \cdot 14 \pmod{17}$
5. $187 \cdot 185 \cdot 4 \pmod{188}$
6. $14 \cdot 15 + 15 \cdot 6 + 16 \cdot 9 \pmod{19}$
7. $83 \cdot 10 + 50 \pmod{99}$
8. $16 \cdot 15 \pmod{32}$
9. $23 \cdot 24 \pmod{47}$

"Did you notice that something rather interesting happened with part 2 above? When multiplying $11 \cdot 15 \pmod{33}$ we rewrote it as $11 \cdot 3 \cdot 5$ which is 0 since $11 \cdot 3 = 0 \pmod{33}$. When we obtain a factor of 0, the computation becomes considerably easier!"

Answer, 2.31
It's still 4.

Answers, 2.32
These are the calculations that occurred to me. You may have found more clever methods than I.

1. $15 \cdot (2 \cdot 6) \cdot 7 \cdot -2 \cdot 8 = (15 \cdot 2) \cdot 6 \cdot 7 \cdot -16 = -1 \cdot 7 \cdot (3 \cdot 2) \cdot 15 = -1 \cdot 7 \cdot 3 \cdot -1 = 21 \pmod{31}$.
2. $(11 \cdot 3) \cdot 5 \cdot 17 \cdot 26 = 11 \cdot 3 \cdot \text{junk} = 0 \pmod{33}$ since $11 \cdot 3 = 0 \pmod{33}$.
3. $17 \cdot 12 + 12 \cdot 12 = 29 \cdot 12 = -1 \cdot 12 = -12 = 18 \pmod{30}$.
4. $-1 \cdot -2 \cdot -3 = -6 = 11 \pmod{17}$.
5. $-1 \cdot -3 \cdot 4 = 12 \pmod{188}$.
6. $(14 + 6)15 + 8(2 \cdot 9) = 1 \cdot 15 + 8 \cdot -1 = 15 + -8 = 7 \pmod{19}$.
7. $-16 \cdot 10 + 5 \cdot 10 = -11 \cdot 10 = -(10 \cdot 10 + 1 \cdot 10) = -(1 + 10) = -11 = 88 \pmod{99}$.
8. $16(2 \cdot 7 + 1) = (16 \cdot 2)7 + 16 = 0 + 16 = 16 \pmod{32}$.
9. $23 \cdot 24 = 22 \cdot 24 + 1 \cdot 24 = 11(2 \cdot 24) + 24 = 11 \cdot 1 + 24 = 35 \pmod{47}$.

"Pretty shrewd, Ant! I wish I'd thought to look for that way back in $\mathbf{Z}_{11}$. It would have made my calculations a lot simpler!"

"You would have been spinning your wheels, Gnam. In mod 11 arithmetic, the only way a product ab can be 0 is if at least one of the factors a or b is 0. That is,

$$\text{If } ab = 0 \pmod{11} \text{ then } a = 0 \text{ or } b = 0.$$

"But this property of $\mathbf{Z}_{11}$ doesn't hold for $\mathbf{Z}_{33}$. It is possible to get a mod 33 product of 0 when neither factor is 0. In other words, $\mathbf{Z}_{33}$ has *zero divisors* whereas $\mathbf{Z}_{11}$ doesn't."

Definition

A nonzero element a of $\mathbf{Z}_n$ is called a **mod n zero divisor** if there exists a nonzero element b of $\mathbf{Z}_n$ such that $ab = 0 \pmod{n}$.

"So, for example, 5 is a mod 15 zero divisor because $5 \cdot 3 = 0 \pmod{15}$. Of course, for the same reason, 3 is a mod 15 zero divisor. But 5 is not a mod 11 zero divisor since the only way $5b$ can be 0 (mod 11) is if $b = 0$."

Exercise 2.33

Verify that $\mathbf{Z}_{11}$ has no zero divisors by checking that your mod 11 multiplication table contains no zeros, except when one of the factors is 0.

Exercise 2.34

1. Does $\mathbf{Z}_{12}$ contain any mod 12 zero divisors? Justify your answer.
2. Does $\mathbf{Z}_7$ contain any mod 7 zero divisors? Justify your answer.
3. Does $\mathbf{Z}_{35}$ contain any mod 35 zero divisors? Justify your answer.

Answers, 2.34

1. Yes. For example, $3 \cdot 4 = 0 \pmod{12}$. So 3 and 4 are both mod 12 zero divisors. (Incidentally, 2, 6, 8, 9, and 10 are also mod 12 zero divisors.)
2. There are no mod 7 zero divisors. One way to see this is to make a mod 7 multiplication table and note that wherever 0 is an entry, one of the factors is 0.
3. Yes. For example $7 \cdot 5 = 0 \pmod{35}$. (The set of mod 35 zero divisors is $\{5, 7, 10, 14, 15, 20, 21, 25, 28, 30\}$. If you multiply 5, 10, 15, 20, 25 or 30 by a multiple of 7 the product is 0. If you multiply any of the other numbers, 7, 14, 21, or 28, by a multiple of 5 the product is 0.)

"I think I know which modular systems have zero divisors and which don't, Ant! It's really all quite simple once you see what's going on, and although I don't mean to brag, the picture is quite clear to me. The key here is the concept pride, I mean prime."

"I believe the time has come to consider some of the basic concepts of number theory such as primes and units and composites. Then we'll return to our study of the modular systems and some of their powerful applications."

"You are planning on ending Section 2, here? But we haven't analyzed the Black Magic Device yet, and you promised we would do so!"

"Patience, Gnam. We'll get to it."

Section Three

Basic Concepts of Factorization and Divisibility

"Ant, you have interrupted a rather interesting point I was about to make about the modular systems to rehash some boring stuff. Please give serious consideration to skipping this section. I don't even know what it is doing in your book. Everyone knows that primes are numbers like 5 which have only two factors, and that the number 1 is or is not a prime depending upon whom you are talking to."

"Well, if what you say is true, then everybody knows wrong, because a prime can have any number of factors depending upon the domain under consideration. Furthermore, 5 is a composite Gaussian integer, but a unit in the context of $\mathbf{Z}_{11}$. Also, 1 is invariably used to denote the multiplicative identity of an algebraic structure and so *ipso facto* is a unit not a prime. So there!"

"Ant, sit down! You're making no sense. I think you're losing it."

"No, Gnam. You sit down, and prepare yourself for a liberating experience. You are about to learn what 'prime' really means. The definitions that we will develop here will apply to *any* integral domain, not just to the set of positive integers (which isn't even an integral domain)."

"There you go again! What's an integral domain?"

"An **integral domain** is a commutative ring with no zero divisors. For example, the entire set of integers is an integral domain, as is $\mathbf{Z}_{11}$ under mod 11 addition and multiplication. But $\mathbf{Z}_{15}$ isn't because it has zero divisors.

"Now, I happen to have a favorite integral domain, but it is bashful and doesn't wish to be identified. So in all the definitions that follow, I'm going to refer to my integral domain as S. (You may think of S as the 'world' temporarily under consideration.)*"

Definition

If a and b are elements of S, then **a is a factor of b in S** if and only if there exists an element c in S such that $ac = b$. "**$a \mid b$**" means "a is a factor of b". "**$a \nmid b$**" means "a is not a factor of b".

Notation

Z refers to the set of integers $\{\ldots, -2, -1, 0, 1, 2, \ldots\}$.

Examples:

Suppose that $S = \mathbf{Z}$. Within this domain, $2 \mid -4$ since $2(-2) = -4$. Also, $3 \mid 12$, $-5 \mid 20$ and $1 \mid -1$. But $2 \nmid 5$ since there is no integer c such that $2c = 5$.

Suppose that $S = \mathbf{Z}_{11}$ under mod 11 multiplication. Within this domain, $2 \mid 5$ since $2 \cdot 8 = 5 \pmod{11}$. But $0 \nmid 5$, since there is no element c of $\mathbf{Z}_{11}$ such that $0 \cdot c = 5 \pmod{11}$.

Let S be the set of rational numbers. Within this domain, $2 \mid 5$ since $2(5/2) = 5$. Also, $(2/3) \mid (5/9)$ since $(2/3)(5/6) = 5/9$.

"Gnam, please don't confuse '$a \mid b$' with 'a/b'. '$a \mid b$' is the assertion that a is a factor of b. 'a/b' means a times the multiplicative inverse of b.

"Now, there are other ways of saying 'a is a factor b', including:

b is a multiple of a

a is a divisor of b

and a divides b (my favorite way).** "

Examples: Let $S = \mathbf{Z}$. Then within this domain, 3 divides 12, –15 is a multiple of –5, and 11 is a divisor of –33. But 17 is not a multiple of 4, and –2 does not divide –5.

---

* For the rest of this section you can give the modular systems $\mathbf{Z}_n$, where n is a composite, a rest. With the exception of the definition on this page, none of the concepts developed here apply to them. However, in order to demonstrate a specific point, I will occasionally apply the concepts developed in this section to subsets of the integers which aren't integral domains.

** Gnam's favorite way is "b is a bunch of a's".

Exercise 3.1

1. Let S be the set of rational numbers. Show that $(1/2) \mid 4$, $7 \mid (5/3)$, and $(2/5) \mid (-3/4)$. Does 0 divide 3?
2. Let $S = \mathbf{Z}_3$. Show that $2 \mid 1$ and $1 \mid 2$, but that $0 \nmid 1$.
3. The tyrant was described as "odd," and in a fit of fury he banished all odd numbers from his kingdom. His subjects had to do all of their mathematics within the set of even integers, $E = \{\ldots, -4, -2, 0, 2, 4, 6, \ldots\}$. (This set is not an integral domain because it lacks a multiplicative identity, but what could the poor people do?) Within this world, does $2 \mid 2$? does $2 \mid 8$? does $4 \mid 12$?

"The moral of the story is that, in order to answer the question of whether one number divides another, we need to know which domain the numbers are living in. However, in any respectable world, 0 never divides anything but itself, and so merits no consideration when discussing concepts of factorization. (Please toss 0 out the window. Thank you.) On the other hand, 1 divides everything in sight including itself. Indeed, numbers such as 1 which are factors of everything are called units."

Definition

An element u of S is called a **unit of S** if and only if for each b in S, $u \mid b$.

Examples:

Let $S = \mathbf{Z}$. The units of $\mathbf{Z}$ are 1 and -1.

Let $S = \mathbf{Z}_3$ under $\cdot_3$. The units of $\mathbf{Z}_3$ are 1 and 2. Clearly $1 \mid 0$, $1 \mid 1$, and $1 \mid 2$. But it is also the case that $2 \mid 0$, $2 \mid 1$ since $2 \cdot 2 = 1 \pmod 3$, and $2 \mid 2$.

Let S be the set of rational numbers. Every rational number except 0 is a unit.

Answers, 3.1

1. $(1/2) \cdot 8 = 4$; $7 \cdot (5/21) = 5/3$; $(2/5) \cdot (-15/8) = -3/4$. No. There is no rational c such that $0 \cdot c = 3$. (Note: If r and s are *any* rational numbers, where $r \neq 0$, then $r \mid s$ since $r(s/r) = s$.)
2. $2 \mid 1$ since $2 \cdot 2 = 1 \pmod 3$. $1 \mid 2$ since $1 \cdot 2 = 2 \pmod 3$. But $0 \nmid 1$ since there is no c such that $0c = 1 \pmod 3$.
3. 2 is not a factor of 2 since there is no c in E such that $2c = 2$ (1 isn't an element of the set); $2 \mid 8$ since $2 \cdot 4 = 8$; $4 \nmid 12$ since there is no c in E such that $4c = 12$.

Consider the set of even integers, E. 2 is not a unit of E since, for example, 2 is not a factor of 6 in E. (There is no k in E such that $2k = 6$.) E has no units.

"Back to our arbitrary integral domain. We've discarded 0. Would you please put all the units in one corner of the room? Thank you. Now, there may be some numbers left which have all sorts of interesting factors. For example, consider the integer 6 in the context of the **positive integers**. 1, 2, 3, and 6 are all factors of 6. The factor 1 is not interesting, since 1 is a factor of every number. And the factor 6 is not interesting, because every integer is a factor of itself. So, $6 = 6 \cdot 1$ is hardly worth mentioning. But $6 = 2 \cdot 3$ is definitely worth noting. Numbers with interesting factorizations are called **composites**."

Definition

A nonzero element c of S is called a **composite of S** if and only if there is at least one factorization of c into a product ab where a and b are elements of S and neither a nor b is a unit of S.

Examples:

Let $S = \mathbf{Z}$. Then -6 is a composite of $\mathbf{Z}$ since $-6 = -2 \cdot 3$, and neither -2 nor 3 is a unit of $\mathbf{Z}$. However, -5 is not a composite, since the only ways to factor -5 are $-5 \cdot 1$ and $5 \cdot (-1)$. In both cases, one of the factors is a unit.

Let $S = \mathbf{Z}_3$ under mod 3 multiplication. We've thrown 0 out the window, and 1 and 2 are both units and hence belong in the northwest corner of this room. There are no numbers left! $\mathbf{Z}_3$ has no composites.

Exercise 3.2

1. Let S be the set of rational numbers. What are the composites of S?
2. What are the composites of E, the set of even integers?
3. Show that in any commutative ring, 1 and -1 are units.

Answers, 3.2

1. There are none. For any pair of rational numbers a and b, $a \neq 0$, $a \mid b$ since $a(b/a) = b$. Thus, all rationals except 0 are units.
2. The composites are the multiples of 4, since $4k = 2(2k)$ and both 2 and $2k$ are nonunits of E.
3. If a is an element of a ring, then $1a = a$, so $1 \mid a$. Also, $(-1)(-a) = a$, so $-1 \mid a$.

"After you have tossed out 0 and set aside the units and the composites, what's left are the primes. They are the nonunits with only boring factorizations. That is, whenever primes are factored into ab, then either a or b, but not both, is a unit."

Definition

An element p of S is **a prime of S** if and only if whenever $p = ab$, where a and b are elements of S, either a or b (not both) is a unit of S.

Examples: Let $S = \mathbf{Z}$. 1 is *not* a prime since the only ways to factor 1 are $1 \cdot 1$ and $-1 \cdot (-1)$. In each case *both* factors are units, not precisely *one*, as required by the definition. For the same reason, -1 is not a prime. However, -3 is a prime, since if $-3 = ab$ then either a or b (but not both) is 1 or -1.

"Notions such as units and primes are normally applied only to integral domains. In any integral domain it is always possible to factor a prime into $p \cdot 1$. However, if we apply these notions to the set of even integers, E (which is 'almost' an integral domain—it lacks the multiplicative identity), the primes have no factorization. For example, 6 is an element of E, and 6 is neither a unit nor a composite. 6 is a prime of E."

Exercise 3.3

1. Let S be the set of rational numbers. Which elements are prime?
2. Let $S = \mathbf{Z}_3$ under mod 3 multiplication. Which elements are prime?
3. Which elements of E, the set of even integers, are prime?

Supplementary Problem 3.1

The Gaussian integers, **G**, are the complex numbers $a + bi$ where a and b are integers. Find the units of **G**. Show that 2 and 5 are not prime, but composite Gaussian integers. Which other prime integers are composite Gaussian integers?

Answers, 3.3

1. and 2. The sets have no primes. All elements except 0 are units.

3. The primes are $\{\pm 2, \pm 6, \pm 10, \pm 14, \ldots\}$. There is *no* way to factor any of these numbers in E.

Supplementary Problem 3.2

Let $R[x]$ be the set of polynomials with real coefficients.

1. Show that the polynomials of degree 0 are the units of $R[x]$.
2. Show that all polynomials of degree 1 are prime polynomials.
3. What conditions on a, b, and c ensure that $ax^2 + bx + c$ is prime? That it is composite?
4. Are there any prime polynomials of degree 3?
5. List all prime polynomials of degree 4 or larger.

"The primes can be thought of as the atoms which make up the composite numbers under multiplication. For example, within the context of the positive integers, 15 is built from the atoms (primes) 3 and 5. That is, $15 = 3 \cdot 5$. You may have seen the procedure illustrated here for finding the prime factors of a positive integer, say 1155. Essentially, you just keep factoring until you are left with nothing but primes. The procedure works well as long as it's clear how to proceed."

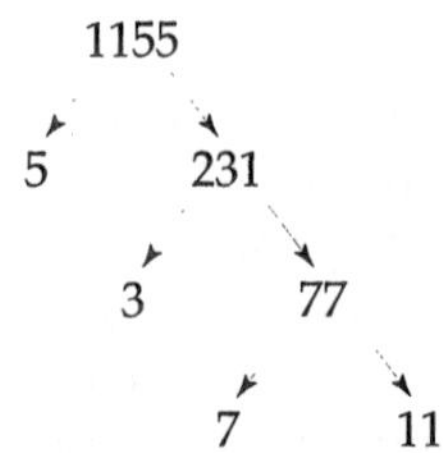

"So, the prime factors of 1155 are 3, 5, 7, and 11. An added benefit of this procedure is that it also provides us with a factorization of 1155 into a product of primes: $1155 = 3 \cdot 5 \cdot 7 \cdot 11$."

Exercise 3.4

Let $S = \mathbf{Z}^+$, the set of positive integers. ($\mathbf{Z}^+$ isn't an integral domain: no 0; no additive inverses; no care!) Factor each of the following into a product of primes: 126; 280; 858.

"Now consider 15 within the context of $\mathbf{Z}$. There are two ways to factor 15 into a product of primes, $3 \cdot 5$ and $(-3)(-5)$. Similarly, -15 has two factorizations into primes: $-3 \cdot 5$ and $3 \cdot (-5)$. A common convention pertaining to $\mathbf{Z}$ is to prefer factorizations which are products of *positive* primes only, tossing in a factor of -1 if

---

Answers, 3.4
$126 = 2 \cdot 3^2 \cdot 7$; $280 = 2^3 \cdot 5 \cdot 7$; $858 = 2 \cdot 3 \cdot 11 \cdot 13$.

necessary. So the answer to 'factor 15 and –15 into products of primes' is $15 = 3 \cdot 5$ and $-15 = -1 \cdot 3 \cdot 5$. Of course, –1 is not a prime, but it is allowed to appear at most once in the factorization."

Exercise 3.5

1. Let $S = \mathbf{Z}$. Factor –28 and –31 into a product of primes.
2. Let E be the set of even integers. Factor 100 into a product of primes of E.

"One of the distinguishing characteristics of **Z** is there is only one way to express a composite as a product of primes (given the convention described above). That is, if both you and I factor a composite into a product of primes and if neither of us makes a mistake, then we will obtain the same answer, except possibly for the order of our factors. This property is called the **Unique Factorization Theorem.*** It doesn't hold in all integral domains.

"The Unique Factorization Theorem for the integers may be the most important result in number theory, and yet until about 150 years ago no one gave it much thought. Indeed, the 19th-century mathematician E. E. Kummer erroneously believed that he had proved Fermat's Last Theorem by using a technique which involved factoring polynomials over subdomains of the complex numbers. No doubt he was incredibly elated, for surely everlasting fame would befall anyone who either proved or found a counterexample to the famous "theorem".** It must have been an enormous letdown when another mathematician, Dirichlet, pointed out that Kummer's argument assumed that all of the integral domains he was working in had the unique factorization property. But most of them didn't! Kummer's "proof" was null and void.

Answers, 3.5

1. $-28 = -1 \cdot 2^2 \cdot 7$; –31 is a prime, but according to the convention we would write $-31 = -1 \cdot 31$.
2. There are *two* ways of doing this: $100 = 10 \cdot 10$ and $100 = 50 \cdot 2$

* Also called the Fundamental Theorem of Arithmetic

** In 1993, it was claimed that Princeton mathematician Andrew Wiles had proved Fermat's Last Theorem, after 300 years of attempts to do so by some of the world's greatest mathematicians. We'll return to this famous conjecture in Section 6.

"In any integral domain every composite is a product of primes. But you might not have *unique* factorization. That is, depending upon the domain, it may be possible to factor elements into products of primes in more than one way.*

"Incidentally, if 1 were considered a prime, as well as a unit, then the unique factorization theorem would no longer be true for the integers. $2 \cdot 3, 1 \cdot 2 \cdot 3$, and $1 \cdot 1 \cdot 2 \cdot 3$ would be different factorizations of 6 into products of primes."

Note: *The rest of this section pertains to the domain of integers.*

"How would you factor a really enormous integer, Ant, such as 212658493627364581?"

"Personally, I wouldn't. There are techniques for sending coded messages which are essentially spy-proof because to break the messages one must be able to factor very large composites. One can program a computer to seek factors, but even modern-day computers have their limits. Nevertheless, there are strategies which are effective on reasonably small numbers, such as 1157.

"First observe that 1157 is not a multiple of 2. Nor is it a multiple of 3. We don't need to check whether 4 is a prime factor since 4 isn't prime. 1157 is not a multiple of 5. 6 doesn't need to be checked. (Why?) 1157 is not a multiple of 7. And 8, 9, and 10 can be skipped. 11 is not a factor of 1157. The next prime is 13, and, hey, 13 divides 1157. In fact, $1157 = 13 \cdot 89$. Now we need only factor 89. Since none of the primes from 2 to 11 is a factor of 1157, none is a factor of 89."

"I get it. So we just check whether 13, 17, 19, *etc.* are factors of 89."

"Actually, that's not necessary. For if $89 = a \cdot b$, then either a or b must be less than or equal to the square root of 89, which is roughly 10. (If a and b were both greater than the square root of 89, their product would be larger than 89.)

"Now, we know that 89 has no prime factors less than 10 since 89 is a factor of 1157 and 1157 has no prime factors less

---

*Actually, this description of unique factorization is simplistic. In a unique factorization domain, you may well have $N = pq$ and $N = rs$ where p, q, r, and s are different primes. However, then p will be a unit times, say, r and q will be a unit times s.

than 10. Thus, 89 is prime and the unique factorization of 1157 into primes is $13 \cdot 89$.

"In general, a positive integer is prime if it has no prime factors less than or equal to its square root. For if $n = ab$, then a and b can't *both* be larger than the square root of n."

"One last question, Ant. I can see how you dispensed with 2, 5, and 11, but how did you figure out so quickly that 3 and 7 aren't factors of 1157, but that 13 is?"

"In Section 5, where we will analyze the Black Magic Device, we'll learn divisibility techniques. Meanwhile, I suggest you use your calculator to solve the next exercise."

Exercise 3.6

Which of the following integers are prime? Factor each composite integer into a product of positive primes.

1. 2311 2. 757 3. 1859

"When you factored 1859, you probably first noted that $1859 = 11 \cdot 169$ and then factored 169. Any divisor of 169 is also a divisor of 1859. In general, if $b \mid c$ and $c \mid d$, then $b \mid d$."

Theorem 3.1

1. If $b \mid c$ and $c \mid d$ then $b \mid d$.
2. If $b \mid c$ and $b \mid d$ then $b \mid mc$ and $b \mid (mc \pm nd)$.

"Part 1 is clear, Ant. Just imagine d as a bunch of c's. Then since each c is a bunch of b's, it follows that d is a bunch of b's."

"You've just provided a proof, though few mathematicians would say it in quite that way!"

"And part 2 asserts that since six is a bunch of threes, any number of sixes is also a bunch of threes. Also, since nine is a bunch of threes then any number of nines added or subtracted from any number of sixes is also a bunch of threes."

"Yes. But your explanation of part 1 is better since it is a general argument. In the next exercise, I will address these and

---

Answers, 3.6

1. 2311 is a prime since it has no prime factor less than or equal to its square root, which is about 48.
2. 757 is prime since it has no prime factor less than or equal to its square root, which is about 27.
3. $1859 = 11 \cdot 13^2$.

other relationships in number theory. Of course, since it's important for you to learn to be skeptical even of my assertions, not all the statements in the list are true!"

Exercise 3.7

Which of the following statements about integers do you agree with, and which do you think are at least sometimes false? Try to explain *why* you agree or disagree.

1. If $10a + 15b = c$ then $5 \mid c$.
2. If $21a + b = 14$ then $7 \mid b$.
3. If $3a + 5b = 301$ then b might possibly be a multiple of 3.
4. If $a \mid b$ and $a \mid c$ then $a \mid (2b + 3c)$.
5. If $a \mid b$ and $a \mid c$ then $a \mid (5b - 7c)$.
6. If $6 \mid ab$ then $6 \mid a$ or $6 \mid b$.
7. If $7 \mid ab$ then $7 \mid a$ or $7 \mid b$.
8. If $6 \mid (12x + 3y)$ then $6 \mid y$.
9. If $7 \mid (35a + 5b)$ then $7 \mid b$.
10. If $10 \mid 21b$ then $10 \mid b$.
11. If $10 \mid 15b$ then $10 \mid b$.
12. If $20 \mid 15b$ then $4 \mid b$.
13. If $15 \mid (30a + 5b)$ then $15 \mid b$.
14. If $7 \mid (21a + 2b)$ then $7 \mid b$.
15. If $5 \mid (3a - 5b)$ then $5 \mid a$.

Answers, 3.7

1. True, since $c = 10a + 15b = 5(2a + 3b)$.
2. True, since $b = 7(2 - 3a)$.
3. False. If $b = 3k$ then $301 = 3a + 15k = 3(a + 5k)$, but 3 is not a factor of 301.
4. True. If $b = na$ and $c = ma$ then $2b + 3c = 2na + 3ma = a(2n + 3m)$. So, $a \mid (2n + 3m)$.
5. True. Reason as in part 4.
6. False. For example, $6 \mid 4 \cdot 3$ but 6 is not a factor of either 4 or 3.
7. True.
8. False. For example, let $x = 1$ and $y = 2$.
9. True. If $7 \mid (35a + 5b)$ then since $7 \mid 35a$, it follows that $7 \mid 5b$. So $7 \mid b$.
10. True.
11. False. For example, let $b = 2$.
12. True.
13. False. For example, let $a = 1$ and $b = 3$.
14. True. If $7 \mid (21a + 2b)$ then $7 \mid 2b$ and so $7 \mid b$.
15. True.

"One of the key ways in which primes differ from composites is illustrated in parts 6 and 7 above. If $b \mid cd$ it doesn't necessarily follow that $b \mid c$ or $b \mid d$. If b is a composite, it can pick up some of its prime factors from c and the rest of its prime factors from d.

"For example, if $b = 6$, then c can be any multiple of 3 and d can be any multiple of 2, and 6 will divide cd (*e.g.* $6 \mid 9 \cdot 10$). But if b is prime, there is no interesting way to split b apart. Either c or d (or both) must be multiples of b. Similarly, in part 14, since $7 \mid (21a + 2b)$, 7 must divide $2b$. 2 contributes nothing. So $7 \mid b$. However in part 13, while it is true that $15 \mid 5b$, we can conclude only that $3 \mid b$. It isn't necessarily true that $15 \mid b$.

"Indeed, proofs of the Unique Factorization Theorem generally depend upon the following theorem, which I will state here without proof."

Theorem 3.2

Suppose that p is prime. If $p \mid ab$ then $p \mid a$ or $p \mid b$.

"A similar idea is illustrated in parts 10, 11, and 12. If $10 \mid 21b$, then since 21 contributes none of the prime factors of 10, it's up to b to carry the whole load. Here's a proof: If $10 \mid 21b$, then there is an integer c such that $10c = 21b$. So, $2 \cdot 5 \cdot c = 3 \cdot 7 \cdot b$. Since the same number is represented on both sides of the equation, and its factorization into primes is unique, it follows that the primes 2 and 5 must be factors of b (and that 3 and 7 are factors of c). Thus $10 \mid b$. But the strongest conclusion that can be drawn from $10 \mid 15b$ is that $2 \mid b$."

"I suppose that this idea generalizes into some sort of theorem, too."

"Yes, but it would be helpful to use the term 'common divisor' in the statement of the theorem."

Definition

If $b \mid c$ and $b \mid d$ then b is called a **common divisor** of c and d.

Examples: 2 is a common divisor of 6 and 8; 4 is a common divisor of 4 and 12; but, 4 is not a common divisor of 6 and 8 since 4 does not divide 6.

Theorem 3.3

If $b \mid cd$, but b and c have no common divisor greater than 1, then $b \mid d$.

Exercise 3.8

Suppose that a, b, and c are integers. For each of the following, what are the strongest conclusions that can be drawn?

1. $4a + 5b = 2.$
2. $9a - 12b = c.$
3. $15a - b = 3c.$
4. $7a + 14b = c.$
5. $10a - 2b = 5c.$
6. $12a + 15b = 10c.$
7. $5a = 7b.$
8. $5 \mid 7b.$
9. $15 \mid 5b.$
10. $7 \mid ab.$
11. $6 \mid 3b.$
12. $15 \nmid (5a + 30).$
13. $12 \mid 4(a + 3b).$
14. $15 \mid (9a + 30).$
15. $24 \mid (16a + 8b).$
16. $8 \nmid (16b + c).$

Exercise 3.9

Use the Unique Factorization Theorem to solve for the integer k (if possible) in each of the following equations.

1. $2^3 \cdot 3 \cdot 5^2 \cdot 7 \cdot k = 2^3 \cdot 3 \cdot 5^3 \cdot 7 \cdot 11.$
2. $3 \cdot 5^2 \cdot 7 = 2 \cdot 3 \cdot 5 \cdot 7 \cdot k.$
3. $11^2 \cdot 13 \cdot k = 2^3 \cdot 3 \cdot 11^3 \cdot 13.$
4. $6 \cdot 15 \cdot k = 2^3 \cdot 3^2 \cdot 5^2.$
5. $6 \cdot 15 \cdot k = 2 \cdot 3 \cdot 5 \cdot 7.$

Answers, 3.8

1. $2 \mid b$ but $4 \nmid b$; also $5 \nmid a$.
2. $3 \mid c$.
3. $3 \mid b$.
4. $7 \mid c$.
5. $5 \mid b$ and $2 \mid c$.
6. $5 \mid a$ and $2 \mid b$ and $3 \mid c$.
7. $5 \mid b$ and $7 \mid a$.
8. $5 \mid b$.
9. $3 \mid b$.
10. $7 \mid a$ or $7 \mid b$.
11. $2 \mid b$.
12. $3 \nmid a$.
13. $3 \mid a$.
14. $5 \mid a$.
15. $3 \mid (2a + b)$.
16. $8 \nmid c$.

Answers, 3.9

1. $k = 5 \cdot 11 = 55.$
2. No solution since 2 is a factor of the right hand side of the equation, but 2 is not a factor of the left hand side.
3. $k = 8 \cdot 3 \cdot 11 = 264.$
4. First factor $6 \cdot 15$ into $2 \cdot 3^2 \cdot 5$. Then $k = 2^2 \cdot 5 = 20.$
5. No solution since 3^2 is a factor of the left hand side of the equation, but it is not a factor of the right hand side.

Supplementary Problem 3.3

Prove that if a, b, c, x, and y are integers then:

1. If $a \mid b$ and $b \mid c$ then $a \mid c$.
2. If $a \mid b$ and $a \mid c$, then $a \mid (bx + cy)$ and $a \mid (bx - cy)$
3. $a \mid b$ iff $ca \mid cb$ $(c \neq 0)$

Supplementary Problem 3.4

Compute the number of positive factors of $N = p_1^{q_1} \cdot p_2^{q_2} \cdot p_3^{q_3} \cdot \ldots \cdot p_k^{q_k}$ where $p_1, p_2, \ldots p_k$ are distinct primes.

(Suggested Strategy: Simplify the problem. First consider the case where N is of the form p^q. For example, count the number of factors of 3^4 and the number of factors of 2^6. Once you can answer the question for numbers which are powers of one prime, then consider numbers which are products of powers of two primes. For example, systematically list and then count the factors of $2^2 \cdot 3^4$. Do a couple more examples like this until you know how many factors $p_1^{q_1} \cdot p_2^{q_2}$ has. Then repeat the process for products of powers of three primes, *etc.* until you can generalize.

Alternately, if you are familiar with counting techniques, there is an easier approach. For example, any factor of, say, $2^4 \cdot 3^2 \cdot 5^7$ has to be of the form $2^r \cdot 3^s \cdot 5^t$ where $0 \leq r \leq 4$, $0 \leq s \leq 2$, and $0 \leq t \leq 7$. How many such factors are there? Now generalize.)

"While we're on the subject of divisibility and factorization, we need to explore the concepts of greatest common divisor and least common multiple, for they will be useful when we analyze the Football Problem and Male Chauvinist Chess in later sections. Then I'll treat you to my favorite proof that $\sqrt{2}$ is not rational. And, finally, I'll top off this section with an anecdote concerning my earliest edifying mathematical experience. (So quit playing Tetris, Gnam!)"

Notation

gcd(c, d) means "the greatest common divisor of c and d".

Examples: gcd(2, 6) = 2; gcd(–12, 16) = 4; gcd(1, 3) = 1; gcd(3, 7) = 1; gcd(–3, –8) = 1.

Definition

If $\gcd(c, d) = 1$, then c and d are called **relatively prime**.

Examples: 2 and 9 are relatively prime, as are 3 and –5, and 6 and 35. In all three cases the greatest common divisor is 1.

Exercise 3.10

Which of the following statements are true and which are false (that is, not always true)? For each false statement, give a counterexample. If the statement is true, illustrate it with a couple of examples.

1. If $b \mid c$ and $b \mid d$ then $b \mid \gcd(c, d)$.
2. If $b \mid c$ and $b \mid d$ then $\gcd(c, d) \mid b$.
3. If either c or d is prime, then c and d are relatively prime.
4. If c and d are both composites, then c and d are not relatively prime.
5. For any integer c, $\gcd(c, 1) = 1$.
6. If $\gcd(b, c) = 1$ and $\gcd(c, d) = 1$ then $\gcd(b, d) = 1$.
7. If $\gcd(b, c) = 6$ and $\gcd(c, d) = 10$, then $3 \nmid d$.
8. If b and c are relatively prime, then b or c is prime.
9. If $a \mid c$ and $b \mid c$ then $ab \mid c$.
10. If $a \mid c$ and $b \mid c$ and $\gcd(a, b) = 1$ then $ab \mid c$.

"There are several ways to find the gcd of two numbers. The one I like the most, the Euclidean algorithm, dates back to the ancient Greeks. To find the greatest common divisor of, say, 72 and 30, you divide 72 by 30 and reason as follows:

$$72 = 2(30) + 12.$$

"Now, any number which divides both 72 and 30 must divide 12 since $72 - 2(30) = 12$. Similarly, any number which divides both 30 and 12 must divide 72 since $2(30) + 12 = 72$. Thus, $\gcd(72, 30) = \gcd(30, 12)$. The problem therefore reduces to finding $\gcd(30, 12)$. If it is obvious to you that $\gcd(30, 12) = 6$, then you can quit. Otherwise, divide 30 by 12 obtaining

$$30 = 2(12) + 6.$$

Answers, 3.10

1.	True.	2.	False.	3.	False.
4.	False.	5.	True.	6.	False.
7.	True.	8.	False.	9.	False.

10. True. (We leave the examples and counterexamples to you.)

"By the same reasoning, gcd(30, 12) = gcd(12, 6). Now, if it isn't evident what gcd(12, 6) is, you divide again:

$$12 = 2(6) + 0.$$

"The remainder 0 means that 6 is a factor of 12. Hence, 6 = gcd(12, 6) = gcd(30, 12) = gcd(72, 30).

"This procedure illustrates the elegant mathematical strategy of repeatedly reducing a problem to easier problems until it becomes evident what the answer is.* To find gcd(72, 30), it suffices to find gcd(30, 12). But to find gcd(30, 12), it suffices to find gcd (12, 6). Now $6 \mid 12$ so gcd(12, 6) = 6. Thus, the original problem is solved.

"Nevertheless, we toiled unnecessarily in finding the quotients. Only the remainders are relevant. We could have argued:

$$r_{30}(72) = 12$$

$$r_{12}(30) = 6$$

and finally

$$r_6(12) = 0.$$

Thus, gcd(72, 30) = 6."

Exercise 3.11

Use the Euclidean Algorithm to find the following.

1. gcd(140, 246)
2. gcd(420, 495)

"A second method of finding the greatest common divisor requires first factoring both numbers into products of primes. To find gcd(72, 30) we reason as follows:

$$72 = 2^3 \cdot 3^2 \text{ and } 30 = 2 \cdot 3 \cdot 5.$$

Hence,

$$\gcd(72, 30) = 2 \cdot 3 = 6.$$

Similarly, $\gcd(2^3 \cdot 3^4 \cdot 5 \cdot 7^2, 3^2 \cdot 5^2 \cdot 7 \cdot 11) = 3^2 \cdot 5 \cdot 7$."

---

*We commonly use this strategy in solving equations. For example, to solve $3(x-2) = 5(2x+3)$, we repeatedly simplify the problem by finding equivalent equations until we obtain an equation of the form x = (something), whose solution is evident.

Answers, 3.11

1. gcd(140, 246) = 2 since: $r_{140}(246) = 106$, $r_{106}(140) = 34$, $r_{34}(106) = 4$, $r_4(34) = 2$, and $r_2(4)=0$.
2. gcd(420,495) = 15 since: $r_{420}(495) = 75$, $r_{75}(420) = 45$, $r_{45}(75) = 30$; $r_{30}(45) = 15$ and $r_{15}(30) = 0$.

Definition

If $a \mid c$ and $b \mid c$ then c is called a **common multiple** of a and b.

Examples: 8 is a common multiple of 4 and 1; 24 is a common multiple of 8 and 12; 1 is a common multiple of 1 and 1.

Notation

lcm(a, b) means "the least common positive multiple of a and b" when $ab \neq 0$. If $ab = 0$ then $\text{lcm}(a, b) = 0$.

Examples: $\text{lcm}(2, 4) = 4$; $\text{lcm}(4, 6) = 12$; $\text{lcm}(-3, 5) = 15$.

"Again, it is fairly simple to find the lcm of two numbers which have been factored into products of primes.

$$63 = 3^2 \cdot 7 \text{ and } 30 = 2 \cdot 3 \cdot 5 \text{ so } \text{lcm}(63, 30) = 2 \cdot 3^2 \cdot 5 \cdot 7 = 630.$$

Similarly,

$$\text{lcm}(2 \cdot 3^3 \cdot 5^2 \cdot 7^5 \cdot 11, 2^4 \cdot 5^3 \cdot 11^2 \cdot 13) = 2^4 \cdot 3^3 \cdot 5^3 \cdot 7^5 \cdot 11^2 \cdot 13."$$

Exercise 3.12

1. Find the gcd and lcm of 36 and 64.
2. Find the gcd and lcm of $2^5 \cdot 3 \cdot 5^3$ and $3^2 \cdot 5^2 \cdot 7$.
3. Find the gcd and lcm of $2^3 \cdot 3 \cdot 5$ and $5^2 \cdot 7 \cdot 11$.
4. Find the gcd and lcm of 31 and 37.
5. Find the gcd and lcm of $6 \cdot 9 \cdot 26$ and $15 \cdot 21$.
6. If p and q are different primes, what is $\gcd(p, q)$ and $\text{lcm}(p, q)$?
7. If b and c are relatively prime, what is $\text{lcm}(b, c)$?
8. What is $(2^2 \cdot 3 \cdot 5 \cdot 7)/(2 \cdot 5 \cdot 7)$?
9. What is $(2 \cdot 3^3 \cdot 5 \cdot 7)/(6 \cdot 21)$?
10. Suppose that $\gcd(a, b) = a$. What is the relationship between a and b?
11. Suppose that $\text{lcm}(a, b) = a$. What is the relationship between a and b?
12. Let $a = 11 \cdot 13 \cdot 17$ and $b = 13 \cdot 19 \cdot 31$. Find the gcd and lcm of a and b. What is $a/\gcd(a, b)$? $b/\gcd(a, b)$? $\text{lcm}(a, b)/a$? $\text{lcm}(a, b)/b$? $\text{lcm}(a, b) \cdot \gcd(a, b)/(ab)$?

13. Suppose that a and b are positive integers and $\gcd(a, b) = 3 \cdot 5^2$ and $\operatorname{lcm}(a, b) = 3^2 \cdot 5^3 \cdot 7$. What is the product ab? In general, write an equation describing the relationship between $a, b, \gcd(a, b)$ and $\operatorname{lcm}(a, b)$.

Definition

Suppose that a and b are positive integers. The fraction a/b is called a **reduced fraction** if and only if $\gcd(a, b) = 1$.

Exercise 3.13

In this exercise, a, b, c, and d are positive integers.

1. Suppose that $\gcd(a, b) = 2$ and $a/b = 3/4$. What is a? b?
2. Suppose that a/b is reduced and $a/b = 3/4$. What is a? b?
3. Suppose that a/b and c/d are both reduced and $a/b = c/d$. What is relationship between a and c? b and d?
4. Suppose that $\gcd(a, b) = 1$, $\gcd(c, d) = 5$ and $a/b = c/d$. What is the relationship between a and c? b and d?
5. Suppose that $\gcd(a, b) = 2$, $\gcd(c, d) = 6$ and $a/b = c/d$. What is the relationship between a and c? b and d?

Answers, 3.12

1. $\gcd = 2^2 = 4$ and $\operatorname{lcm} = 2^6 \cdot 3^2 = 576$.
2. $\gcd = 3 \cdot 5^2$ and $\operatorname{lcm} = 2^5 \cdot 3^2 \cdot 5^3 \cdot 7$.
3. $\gcd = 5$ and $\operatorname{lcm} = 2^3 \cdot 3 \cdot 5^2 \cdot 7 \cdot 11$.
4. $\gcd = 1$ and $\operatorname{lcm} = (31)(37) = 1147$.
5. $\gcd = 3^2$ and $\operatorname{lcm} = 2^2 \cdot 3^3 \cdot 5 \cdot 7 \cdot 13$.
6. $\gcd(p, q) = 1$ and $\operatorname{lcm}(p, q) = pq$.
7. $\operatorname{lcm}(b, c) = bc$.
8. $2 \cdot 3$ or 6.
9. $3 \cdot 5$ or 15.
10. $a \mid b$.
11. $b \mid a$.
12. $\gcd = 13$ and $\operatorname{lcm} = 11 \cdot 13 \cdot 17 \cdot 19 \cdot 31$. $a/\gcd(a, b) = 11 \cdot 17$ and $b/\gcd(a, b) = 19 \cdot 31$. $\operatorname{lcm}(a, b)/a = 19 \cdot 31$; $\operatorname{lcm}(a, b)/b = 11 \cdot 17$. $\operatorname{lcm}(a, b) \cdot \gcd(a, b)/ab = 1$.
13. $ab = 3^3 \cdot 5^5 \cdot 7$. $ab = \operatorname{lcm}(a, b) \cdot \gcd(a, b)$.

Answers, 3.13

1. $a = 6$ and $b = 8$.
2. $a = 3$ and $b = 4$.
3. $a = c$ and $b = d$.
4. $c = 5a$ and $d = 5b$.
5. $c = 3a$ and $d = 3b$.

Exercise 3.14

> Convince yourself that if the fraction a/b is reduced then the fraction a^2/b^2 is also reduced. (That is, that if $\gcd(a, b) = 1$, then $\gcd(a^2, b^2) = 1$.)

"You might be convinced by examining several examples of reduced fractions and observing that in each case their squares are also reduced. Indeed, much of our knowledge comes from considering lots of individual cases and then generalizing our observations. But we should consider such knowledge tentative, for it is possible that we neglected to examine individual instances which would provide counterexamples to our generalization.

"A more compelling argument is a general one; that is, an argument which considers all possible cases. For example, to prove a^2/b^2 is reduced whenever a/b is reduced, we might reason as follows."

Suppose that a/b is a reduced fraction. Then $\gcd(a, b) = 1$. Now either a^2/b^2 is reduced or a^2/b^2 is *not* reduced.

Suppose that a^2/b^2 is *not* reduced. Then a^2 and b^2 have a common factor which is larger than 1. So a^2 and b^2 have a common prime factor. Call one such common prime factor p.

Thus, we have $p \mid (a \cdot a)$ and $p \mid (b \cdot b)$. But that means $p \mid a$ and $p \mid b$. This contradicts our assumption that a and b are relatively prime.

Since we have eliminated the possibility that a^2/b^2 is not a reduced fraction, it follows that a^2/b^2 is a reduced fraction.

"We have just proved the following theorem."

Theorem 3.4

> Suppose that a and b are positive integers. If the fraction a/b is reduced then the fraction a^2/b^2 is also reduced.

Supplementary Problem 3.5

> Prove that if a/b is a reduced fraction, then a^3/b^3 is also reduced.

Answer, 3.14
See discussion immediately following the exercise.

"There is a very elegant application of this theorem. The ancient Greeks proved that $\sqrt{2}$ is not a rational number*; that is, that $\sqrt{2}$ cannot be expressed as a quotient of integers. There are many ways to prove this, but my favorite one uses the theorem we have just established. It goes like this."

Either $\sqrt{2}$ can be expressed as a quotient of positive integers or $\sqrt{2}$ cannot be expressed as a quotient of positive integers.

Suppose that $\sqrt{2}$ can be expressed as a quotient of positive integers. Then, in particular, it can be expressed as a reduced fraction. That would mean that there are relatively prime positive integers a and b such that

$$\sqrt{2} = \frac{a}{b}.$$

Squaring both sides we have

$$\frac{2}{1} = \frac{a^2}{b^2}.$$

Now, since a/b is a reduced fraction, it follows that a^2/b^2 is also a reduced fraction. If a^2/b^2 is a reduced fraction, and $a^2/b^2 = 2/1$, then it follows that $a^2 = 2$ and $b^2 = 1$, since $2/1$ is also a reduced fraction. However, a and b are positive integers and there is no positive integer a such that $a^2 = 2$.

So we have eliminated the possibility that $\sqrt{2}$ can be expressed as a reduced fraction, and hence $\sqrt{2}$ is not a rational number.

"We have just proved the following theorem."

Theorem 3.5

$\sqrt{2}$ is not a rational number.

"What I like most about this argument is that it can be extended easily to numbers other than 2, and to roots other than the square root."

---

*This caused quite a stir at the time, for it violated the Greek's religious belief that 'all' could be expressed with just the numbers 1, 2, 3, *etc.* and their ratios.

Supplementary Problem 3.6

1. Modify the above argument to prove that $\sqrt{3}$ is not rational.
2. Modify the argument to prove that $\sqrt{3/4}$ is not rational.
3. Modify the argument to prove that $\sqrt{5/7}$ is not rational.
4. Modify the argument to prove that $\sqrt{12/36}$ is not rational. (Careful: If $a^2/b^2 = 12/36$ where $\gcd(a^2, b^2) = 1$, then it does not follow that $a^2 = 12$ and $b^2 = 36$.)
5. Where does the argument fall apart if one were to (unwisely) try to apply it to show that $\sqrt{9/16}$ is not rational? that $\sqrt{2/8}$ is not rational?
6. Modify the argument to prove that $\sqrt[3]{15/8}$ is not rational; that $(\sqrt[3]{7/15})^2$ is not rational.

Exercise 3.15

Where in the standard elementary school math curriculum do the concepts of greatest common divisor and least common multiple arise?

"Many of the previous results were derived through the use of the Unique Factorization Theorem. Nevertheless, it wasn't the Unique Factorization Theorem that got me hooked onto number theory, but rather another result.

"When I was in the eleventh grade, my math teacher made the claim that there were infinitely many primes. At first I was skeptical that such knowledge was within the reach of us mortals. There didn't seem to be any pattern to the primes. So, how could human beings *ever* know for sure whether the primes went on forever? I mean, there is no way we could ever list all the integers and then check off that infinitely many of them are prime!

"Then my teacher presented the following argument. (It's similar to the one that Euclid used more than 2000 years ago.) He said, 'Let's try assuming that there are only a finite number of primes, say, three of them. Then we could list all of them: p, q, and r. I will now convince you that the list isn't complete.' "

"Even I can do that, Ant! 2, 3, 5, 7. There I've done it! There are more than three primes."

"I too was somewhat baffled, but I decided to keep an open mind and try to follow. Anyway, he continued, 'Consider the integer $pqr + 1$. It has a prime factor. But p is not a factor of $pqr + 1$, q is not a factor of $pqr + 1$, and r is not a factor of $pqr + 1$. So there must be a fourth prime which *is* a factor of $pqr + 1$.' "

"Hold it, Ant! Why can't p be a factor of $pqr + 1$."

"Because $r_p(pqr + 1) = r_p(pqr) +_p r_p(1) = 1$. Ditto with q and r."

"So how do you go from there having to be four primes, to there having to be infinitely many of them, Ant. I mean, that's a fairly large jump!"

"The point is that the argument generalizes to any finite number. There can't be only 100012 primes because if there were, you could multiply them together and then add 1. The resulting number, K, wouldn't have any one of the 100012 primes as factors, and hence there would have to be a 100013th prime which would be a factor of K. In general, if n is any positive integer, there can't be only n primes, because if there were then you could multiply them together and add 1, obtaining some number K. None of the n primes would be factors of K. But K has to have a prime factor. So there are more than n primes."

"I sort of see, Ant. In other words, no matter how many primes you have, there is always at least one more."

"Exactly. Hence, there are infinitely many primes! Gnam, I was so awed by this simple argument, by the fact that human beings were endowed with the ability to use reason to discover facts that naively would seem to be beyond our capacity to know, that I became committed to a lifelong avocation as an Amateur Number Theorist."

"Pretty heavy, Ant! Pretty heavy! I think that a fitting end to this section would be a statement of the theorem you've just proved."

Theorem 3.6

There are infinitely many primes.

Section Four

The Algebraic Structure of the Modular Systems

"Let's see. I think we left Section 2 with the observation that some modular systems, such as $\mathbf{Z}_{33}$, have zero divisors and others, such as $\mathbf{Z}_{11}$, don't. I may have dozed through parts of Section 3, but the only relevant thing that I recall learning there is that you get angry when people define primes as numbers with precisely two factors or say that one can never tell about one."

"You were going to make a conjecture concerning which modular systems have zero divisors, Gnam. Do you remember what it was?"

"Of course! Let's see now. $\mathbf{Z}_{33}$ has zero divisors since $3 \cdot 11 = 0 \pmod{33}$. So do $\mathbf{Z}_{15}$ and $\mathbf{Z}_{12}$, since $3 \cdot 5 = 0 \pmod{15}$ and $3 \cdot 4 = 0 \pmod{12}$ Yes, it's all perfectly clear. $\mathbf{Z}_n$ has zero divisors when n is a composite. That's because n has what you call 'an interesting factorization', and that factorization has a mod n product of 0."

"In other words, Gnam, your argument goes something like this:

"Suppose n is a composite. Then there exist positive integers a and b which are less than n such that

$$ab = n.$$

Hence $$r(ab) = r(n) \pmod{n},$$

so $$r(a) \cdot r(b) = 0 \pmod{n}.$$

But a and b are elements of $\mathbf{Z}_n$, so $r(a) = a$ and $r(b) = b \pmod{n}$. Thus $a \cdot b = 0 \pmod{n}$ but neither a nor b is 0.

"You have just proved the following theorem."

Theorem 4.1

If n is a composite, then $\mathbf{Z}_n$ has zero divisors.

"It's even evident what the zero divisors are, Ant. For example, the mod 24 zero divisors are the factors of 24: 2, 3, 4, 6, 8, and 12."

"But 15 is also a mod 24 zero divisor, Gnam, since

$$8 \cdot 15 = (8 \cdot 3) \cdot 5 = 0 \cdot 5 = 0 \pmod{24}.$$

"Now what other mod 24 zero divisors did you miss?"

"Who knows! Where did the 8 come from? The trouble with you mathematicians is that you conjure up numbers and then expect everyone else to do the same. We magicians never make that mistake. Indeed, I have never heard of a magician who, after pulling a rabbit out of her hat, has turned to her audience and said 'Okay, now you all do it.' "

"That may be true. But unlike magicians, we mathematicians are willing to share our tricks. I picked the number 8 to multiply by 15 since $24 = 8 \cdot 3$ and 15 supplies the 3. I could also have multiplied 15 by 16 obtaining:

$$16 \cdot 15 = 2 \cdot (8 \cdot 3) \cdot 5 = 2 \cdot 0 \cdot 5 = 0 \pmod{24}.$$

"Here's another example: I will show that 18 is a mod 24 zero divisor. $24 = 4 \cdot 6$. 18 supplies the 6 so I just need to supply the 4.

$$4 \cdot 18 = (4 \cdot 6) \cdot 3 = 0 \cdot 3 = 0 \pmod{24}.$$

"Indeed, I could have chosen any one of several other multiples of 4 to demonstrate that 18 is a mod 24 zero divisor. For example:

$$12 \cdot 18 = 4 \cdot (3 \cdot 6) \cdot 3 = 4 \cdot 0 \cdot 3 = 0 \pmod{24}.$$

"*Now* do you see which other elements of $\mathbf{Z}_{24}$ are mod 24 zero divisors?"

"Why are you picking on me? Am I the only one capable of providing you with answers? Our readers should also have an opportunity to work on this question."

Exercise 4.1

1. List all the mod 24 zero divisors and justify your response.
2. List all the mod 18 zero divisors and justify your response.
3. List all the mod 21 zero divisors and justify your response.
4. List all the mod 7 zero divisors and justify your response.
5. In general, if n is a composite, which elements of $\mathbf{Z}_n$ are zero divisors?
6. If n is a prime, which elements of $\mathbf{Z}_n$ are mod n zero divisors?

Theorem 4.2

Suppose that b is a nonzero element of $\mathbf{Z}_n$. Then b is a mod n zero divisor if and only if $\gcd(b, n) > 1$. In particular, if n is prime then $\mathbf{Z}_n$ has no mod n zero divisors.

Supplementary Problem 4.1

Prove Theorem 4.2.

Supplementary Problem 4.2

Suppose that b is a nonzero element of $\mathbf{Z}_n$ and $\gcd(b,n) > 1$. Find all x in $\mathbf{Z}_n$ such that $bx = 0 \pmod{n}$.

"Enough theory! Let's solve some equations, Ant."

Answers, 4.1

1. $\{2, 3, 4, 6, 8, 9, 10, 12, 14, 15, 16, 18, 20, 21, 22\}$. Example of justification: 21 and 8 are both mod 24 zero divisors since $8 \cdot 21 = (8 \cdot 3) \cdot 7 = 0 \pmod{24}$.
2. $\{2, 3, 4, 6, 8, 9, 10, 12, 14, 15, 16\}$.
3. $\{3, 6, 7, 9, 12, 14, 15, 18\}$.
4. There are none. If $ab = 0 \pmod 7$, then $a = 0$ or $b = 0$.
5. (Did you get this one? Is it hard to express? Here's a way of saying it.) If b is a nonzero element of $\mathbf{Z}_n$, then b is a mod n zero divisor $\Leftrightarrow \gcd(b, n) > 1$, that is, if and only if b and n are not relatively prime.
6. There are none.

Exercise 4.2

Find all elements of $\mathbf{Z}_9$ which are solutions of the following equations.

1. $3x + 4 = 1 \pmod 9$
2. $2x + 5 = 2 \pmod 9$
3. $3x = 3 \pmod 9$

"No problem, Ant. The first equation is just $3x = -3$, so $x = -1$; the second equation is equivalent to $2x = -3$. Clearly nothing works. And obviously 1 is the solution of the third equation."

"Big problem, Gnam. -1 (that is, 8), is not the only solution of the first equation. 2 and 5 are also solutions, since $3(2) + 4 = 1 \pmod 9$ and $3(5) + 4 = 1 \pmod 9$. Equation 2 does have a solution, namely 3 since $2(3) + 5 = 2 \pmod 9$. (You might have seen this had you rewritten $2x = -3$ as $2x = 6$.) Furthermore, equation 3 has solutions 4 and 7, in addition to 1 since $3 \cdot 4 = 3 \pmod 9$ and $3 \cdot 7 = 3 \pmod 9$. Now try your hand at these."

Exercise 4.3

Solve each of the following equations over $\mathbf{Z}_9$.

1. $3x = 5 \pmod 9$
2. $2(x - 5) = 1 - x \pmod 9$
3. $6(x - 1) = 1 \pmod 9$

"It's rather odd, Ant! I was certain that each of these equations had all sorts of solutions, but I plugged in every element of $\mathbf{Z}_9$ and now I'm equally certain that none have any!"

"You're right. They don't. To understand when this can happen, let's reexamine how we solve equations of the form $ax = b$ in ordinary algebra, and focus on which number properties we are using when we do so. Consider the equation $2x = 5$ over the real number system. We might argue as follows."

Answers, 4.2 and 4.3
See the conversations between Gnam and Ant following each exercise.

$$
\begin{aligned}
& & 2x &= 5 & \\
&\Leftrightarrow & \frac{1}{2}(2x) &= \frac{1}{2}(5) & \text{[multiply both sides by 1/2]} \\
&\Leftrightarrow & (\frac{1}{2}\cdot 2)x &= \frac{5}{2} & \text{[Associative Property]} \\
&\Leftrightarrow & 1x &= \frac{5}{2} & \\
&\Leftrightarrow & x &= \frac{5}{2} & \text{[Identity Property of 1].}
\end{aligned}
$$

"So the unique solution to the equation $2x = 5$ over the reals is 5/2. To deduce this, we used the Associative Property of Multiplication, the Identity Property of 1, and the fact that 2 has a multiplicative inverse (namely, 1/2).*

"Now all of the modular systems are commutative rings, and so the associative and identity properties hold. But what about multiplicative inverses? Do all of the elements of, say, $\mathbf{Z}_9$ have multiplicative inverses?"

"Uh..."

"We'll return to that question in the next exercise, Gnam. First I have a thing or two to do."

Definition

> If $ab = 1 \pmod{n}$ then a and b are called **mod n multiplicative inverses**.

Supplementary Problem 4.3

> Prove that an element of a commutative ring can have at most one multiplicative inverse. That is, prove that if a, b, and b' are elements of a commutative ring and $ab = ba = 1$ and $ab' = b'a = 1$ then $b = b'$. (Hint: What is $b'ab$?)

Notation

> "**/a**" is an abbreviation for "the multiplicative inverse of a."
>
> Examples: $/8 = 2 \pmod{15}$; $/8 = 3 \pmod{23}$; $/214 = 214 \pmod{215}$.

---

*We also used the arithmetic fact $(1/2) \cdot 5 = 5/2$.

Exercise 4.4

> Which elements of $\mathbf{Z}_9$ have mod 9 multiplicative inverses, and what are their inverses? Which elements don't have mod 9 multiplicative inverses?

"Now, equation 2 in Exercise 4.2 reduces to $2x = 6 \pmod 9$."

$$\begin{aligned} & 2x = 6 \quad \pmod 9 \\ \Leftrightarrow\ & 5(2x) = 5(6) \pmod 9 \quad \text{[multiply both sides by /2, which is 5]} \\ \Leftrightarrow\ & x = 3 \quad \pmod 9. \end{aligned}$$

"Indeed, in any commutative ring, the equation $ax = b$ has a unique solution when a has a multiplicative inverse. There is *at most* one solution since if $ax = b$, then $/a \cdot (ax) = /a \cdot b$. So $(/a \cdot a)x = b/a$. Thus, $x = b/a$. Furthermore, b/a is a solution since $a(b/a) = a(/a \cdot b) = (a \cdot /a)b = 1b = b$."

Theorem 4.3

> Suppose that a and b are elements of $\mathbf{Z}_n$ and that a has a mod n multiplicative inverse. Then there is precisely one x in $\mathbf{Z}_n$ such that $ax = b \pmod n$. In particular, $x = b/a \pmod n$.

Exercise 4.5

> 1. Which elements of $\mathbf{Z}_{10}$ have mod 10 multiplicative inverses? What are their inverses? Which elements of $\mathbf{Z}_{10}$ don't have mod 10 multiplicative inverses?

Answers, 4.4

1, 2, 4, 5, 7, and 8 have multiplicative inverses. 0, 3, and 6 do not. 3 and 6 are the mod 9 zero divisors. $/1 = 1$, $/2 = 5$, $/4 = 7$, $/5 = 2$, $/7 = 4$, and $/8 = 8 \pmod 9$.

Answers, 4.5

1. Clearly 0 has no multiplicative inverse. 2 has no multiplicative inverse since the multiples of 2 on a 10-hour clock are $1 \cdot 2 = 2, 2 \cdot 2 = 4, 3 \cdot 2 = 6, 4 \cdot 2 = 8$, and $5 \cdot 2 = 0$. (Once you hit 0, the cycle starts over again.) So no multiple of 2 is 1 on a 10-hour clock. This means that if you start at 0 and make jumps of size 2, you will never land on 1. Hence, if you make jumps of size 4 you can't land on 1, since jumping 4 is just jumping 2 twice. So 4 has no mod 10 multiplicative inverse. By similar reasoning, neither do 6 or 8. 5 doesn't, since the mod 10 multiples of 5 are 0 and 5. So the elements of $\mathbf{Z}_{10}$ without mod 10 multiplicative inverses are 0, 2, 4, 5, 6, and 8. All but 0 are mod 10 zero divisors. The elements that are left do have multiplicative inverses. $/1 = 1$, $/3 = 7$, $/7 = 3$, and $/9 = 9$.

2. Answer the same questions for $\mathbf{Z}_{12}$, $\mathbf{Z}_7$, and $\mathbf{Z}_{20}$. What's going on?
3. Solve the following equations over the indicated mod. In each equation the coefficient of x has a multiplicative inverse, so the equation has precisely one solution.

a. $3x = 5 \pmod{10}$ b. $11x = 3 \pmod{12}$
c. $5x = 2 \pmod 7$ d. $3x = 9 \pmod{20}$
e. $11x = 4 \pmod{20}$ f. $5x = 10 \pmod{12}$
g. $2x = 6 \pmod 7$ h. $9x = 2 \pmod{10}$
i. $13x = 4 \pmod{20}$ j. $7x = 6 \pmod{12}$

Supplementary Problem 4.4

Prove that if b has a mod n multiplicative inverse, then b is not a mod n zero divisor. (Hint: Suppose that $cb = 0 \pmod n$. What is $cb/b \pmod n$?)

"Okay, Ant. I see the picture! There are two types of numbers, those which are relatively prime to the mod and those which aren't. The ones that are relatively prime have multiplicative inverses, and the ones that aren't are zero divisors. That is probably true for all mods, so if you want to, you can state some kind of theorem right now."

Answers, 4.5 (cont.)

2. In $\mathbf{Z}_{12}$: $/1 = 1$, $/5 = 5$, $/7 = 7$ and $/11 = 11$. (Of course $/11 = 11$ since $11 = -1 \pmod{12}$, and $(-1)^2 = 1$ in any ring.) The other nonzero elements of $\mathbf{Z}_{12}$ are mod 12 zero divisors and have no multiplicative inverses. In $\mathbf{Z}_7$: every nonzero element has a mod 7 multiplicative inverse. In particular, $/1 = 1$, $/2 = 4$, $/3 = 5$, $/4 = 2$, $/5 = 3$, and $/6 = 6$. In $\mathbf{Z}_{20}$: $/1 = 1$, $/3 = 7$, $/7 = 3$, $/9 = 9$, $/11 = 11$, $/13 = 17$, $/17 = 13$, and $/19 = 19$. (If you had difficulty finding multiplicative in verses, that's okay. We'll discuss techniques for finding inverses later in this section.) The other nonzero elements of $\mathbf{Z}_{20}$ are all zero divisors and have no mod 20 multiplicative inverses.
3. a. 5; $/3 = 7$, so $x = 7(5) = 5 \pmod{10}$.
   b. 9; $/11 = 11$, so $x = 11(3) = 9 \pmod{12}$.
   c. 6; $/5 = 3$, so $x = 3(2) = 6 \pmod 7$.
   d. 3; $/3 = 7$, so $x = 7(9) = 3 \pmod{20}$. (You might have "eyeballed" this answer.)
   e. 4; $/11 = 11$, so $x = 11(4) = 4 \pmod{20}$.
   f. 2; if $5x = 10 \pmod{12}$ then $x = 2$.
   g. 3; if $2x = 6 \pmod 7$ then $x = 3$.
   h. 8; $/9 = 9$, so $x = 9(2) = 8 \pmod{10}$.
   i. 8; $/13 = 17$, so $x = 17(4) = 8 \pmod{20}$.
   j. 6; $/7 = 7$, so $x = 7(6) = 6 \pmod{12}$.

Theorem 4.4

> Suppose that a is a nonzero element of $\mathbf{Z}_n$. Then a has a mod n multiplicative inverse if and only if $\gcd(a, n) = 1$. In particular, if n is prime then each nonzero element of $\mathbf{Z}_n$ has a multiplicative inverse.*

"Now, all of the other equations in Exercises 4.2 and 4.3 are equivalent to equations of the form $3x$ = some number (mod 9) or $6x$ = some number (mod 9) and since neither 3 nor 6 has a mod 9 multiplicative inverse, the strategy for solving linear equations falls apart!"

"So, there's nothing to do but plug in every number to see if it works?"

"One of the great advantages of a finite system is that it is possible (at least theoretically) to test each number to see if it is a solution. But it really isn't necessary here. For example, consider

$$3x = \text{some number} \pmod 9.$$

"The mod 9 multiples of 3 are 0, 3, and 6. Imagine starting at 0 on a 9-hour clock and making jumps of 3. You move to 3, then 6, then 0, then back to 3, *etc.* So if the "some number" isn't 0, 3,

---

*Here is an outline of an argument one could use to show that every nonzero element of $\mathbf{Z}_p$ (p prime) has a mod p multiplicative inverse. (For purposes of this argument, assume that p is at least 5.)

$/2$ exists since: $r_2(p) = 1$. So $p = 2k + 1$. Thus $p + 1 = 2k + 2 = 2(k + 1)$. Therefore, $1 = 2(k+1) \pmod p$. So $/2 = k + 1$.

$/3$ exists since: $r_3(p) = 2$ or $r_3(p) = 1$. If $r_3(p) = 2$ then $p = 3k + 2$, so $p + 1 = 3k + 3 = 3(k + 1)$. So $1 = 3(k + 1) \pmod p$ and $/3 = k + 1$. If $r_3(p) = 1$ then $p = 3k + 1$, so $p + 2 = 3(k + 1)$. Hence, $2 = 3(k+1) \pmod p$. We have already shown that $/2$ exists. Therefore $1 = 3[(k+1)/2]$ and $/3 = (k+1)/2$.

$/4$ exists since: $r_4(p)$ is 3 or 2 or 1. If $r_4(p) = 3$, then $p = 4k + 3$. $p + 1 = 4k + 4$. So $1 = 4(k + 1) \pmod p$ and $/4 = k + 1$. If $r_4(p) = 2$, then $p = 4k + 2$, so $p + 2 = 4(k + 1)$. Thus, $2 = 4(k + 1) \pmod p$. We have already shown that $/2$ exists. So, $1 = 4[(k + 1)/2] \pmod p$ and $/4 = (k + 1)/2$. If $r_4(p) = 1$ then $p = 4k + 1$ so $p + 3 = 4(k + 3)$. $3 = 4(k+3) \pmod p$. We have already shown that $/3$ exists, so $1 = 4[(k+1)/3]$. So, $/4 = (k+1)/3$.

By generalizing this argument, we can show that every nonzero element of $\mathbf{Z}_p$ has a mod p multiplicative inverse. This type of argument is called an inductive argument. We show that $/j$ exists by using the existence of $/i$ for all nonzero i less than j.

A similar, only slightly more complicated, inductive argument can be used to show that all elements of composite mods $\mathbf{Z}_n$ which are relatively prime to n have mod n multiplicative inverses.

or 6 then you can be sure that the equation has no solutions. And if the number is 0, 3, or 6, then you can be certain there will be solutions (actually more than one). The same analysis holds for any equation of the form $6x$ = some number (mod 9), since the mod 9 multiples of 6 are also 0, 3, and 6. (Check this.)"

"Okay. So if I think there are solutions, there's nothing to do but plug in every number to see if it works."

"That's not really necessary, Gnam. Consider the equation $6x = 3 \pmod 9$."

$$\begin{aligned} & 6x = 3 \pmod 9 \\ \Leftrightarrow \quad & 6x - 3 = 0 \pmod 9 \\ \Leftrightarrow \quad & 3(2x-1) = 0 \pmod 9 \end{aligned}$$

which is of the form

$$3 \cdot \text{junk} = 0 \pmod 9.$$

"Now 'junk' *could* be 0, but it doesn't *have* to be 0. Since 3 is a mod 9 zero divisor, there are other possibilities. $3 \cdot 3 = 0$ and $3 \cdot 6 = 0 \pmod 9$. So, you can conclude that junk = 0 or junk = 3 or junk = 6. That is:

$$2x - 1 = 0 \pmod 9 \text{ in which case } x = 5,$$

$$2x - 1 = 3 \pmod 9 \text{ in which case } x = 2,$$

or

$$2x - 1 = 6 \pmod 9 \text{ in which case } x = 8.$$

"So the solutions of $6x = 3 \pmod 9$ are 2, 5, and 8.

"Or, as another example, consider the equation $6x = 6$ (mod 9). 6 has no mod 9 multiplicative inverse, so I'm going to reason as follows."

$$\begin{aligned} & 6x = 6 && \pmod 9 \\ \Leftrightarrow \quad & 6x - 6 = 0 && \pmod 9 \\ \Leftrightarrow \quad & 6(x-1) = 0 && \pmod 9 \\ \Leftrightarrow \quad & x - 1 = 0 \text{ or } x - 1 = 3 \text{ or } x - 1 = 6 && \pmod 9 \\ \Leftrightarrow \quad & x = 1 \text{ or } x = 4 \text{ or } x = 7 && \pmod 9. \end{aligned}$$

"Alternately, I could have observed that 1 is obviously a solution of $6x = 6 \pmod 9$. Furthermore, every three moves of 6 on a 9-hour clock returns me to where I started. So, since $6 \cdot 1 = 6$, it follows that $6 \cdot (1 + 3) = 6$ and $6 \cdot (1 + 6) = 6$. So, 4 and 7 are the other solutions."

Exercise 4.6

Find all solutions of each of the following equations over the indicated mod.

1. $5x = 8 \pmod{10}$
2. $5x = 8 \pmod{9}$
3. $5x = 10 \pmod{15}$
4. $4x = 3 \pmod{9}$
5. $5x = 15 \pmod{20}$
6. $7x = 3 \pmod{13}$
7. $7x = 3 \pmod{14}$
8. $7x = 7 \pmod{14}$

"The procedure I demonstrated for solving linear equations with multiple solutions sometimes is tedious to apply. For example, consider the equation $9x = 3 \pmod{12}$:

$$
\begin{aligned}
& & 9x = 3 \pmod{12} \\
&\Leftrightarrow & 9x - 3 = 0 \pmod{12} \\
&\Leftrightarrow & 3(3x - 1) = 0 \pmod{12},
\end{aligned}
$$

which is true if and only if $3x - 1 = 0 \pmod{12}$, $3x - 1 = 4 \pmod{12}$, or $3x - 1 = 8 \pmod{12}$.

"Now, the first two equations have no mod 12 solutions since there is no x such that $3x = 1 \pmod{12}$ or $3x = 5 \pmod{12}$. (The multiples of 3 on a 12-hour clock are 0, 3, 6, and 9.) To solve $3x = 9 \pmod{12}$, we can go through this business again:

Answers, 4.6

1. The mod 10 multiples of 5 are 0 and 5. No multiple of 5 equals 8. Therefore this equation has no solutions.
2. $/5 = 2 \pmod{9}$. So: $5x = 8 \Leftrightarrow 2(5x) = 2(8) \Leftrightarrow x = 7 \pmod{9}$.
3. The solutions are 2, 5, 8, 11, and 14. $5x = 10 \pmod{15} \Leftrightarrow 5(x-2) = 0 \pmod{15} \Leftrightarrow x - 2$ is 0, 3, 6, 9 or 12. If $x - 2 = 0 \pmod{15}$ then $x = 2$. If $x - 2 = 3 \pmod{15}$ then $x = 5$. Continue in this fashion to obtain the other solutions. Alternately, you can observe that every three moves of 5 on a 15-hour clock returns you to where you started. So, since $2 \cdot 5 = 10$, it follows that $5 \cdot 5 = 10$, $8 \cdot 5 = 10$, $11 \cdot 5 = 10$ and $14 \cdot 5 = 10 \pmod{15}$.
4. $/4 = 7 \pmod{9}$. So, $4x = 3 \Leftrightarrow 7(4x) = 7(3) \Leftrightarrow x = 3 \pmod{9}$.
5. The solutions are 3, 7, 11 ,15, and 19. $5x = 15 \pmod{20} \Leftrightarrow 5(x-3) = 0 \pmod{20} \Leftrightarrow x - 3 = 0$ or $x - 3 = 4$ or $x - 3 = 8$ or $x - 3 = 12$ or $x - 3 = 16$. Alternately, you could reason that it is evident that 3 is one solution of $5x = 15 \pmod{20}$. Furthermore, every four moves of 5 on a 20-hour clock returns you to where you started. Therefore, the other solutions are 7, 11, 15 and 19.
6. The solution is 6. $/7 = 2 \pmod{13}$. Hence $7x = 3 \Leftrightarrow 2(7x) = 2(3) \Leftrightarrow x = 6 \pmod{13}$.
7. No solutions since 3 is not a mod 14 multiple of 7.
8. The solutions are 1, 3, 5, 7, 9, 11, and 13.

$$\begin{aligned} & 3x = 9 && \pmod{12} \\ \Leftrightarrow\quad & 3(x-3) = 0 && \pmod{12} \\ \Leftrightarrow\quad & x-3=0 \text{ or } x-3=4 \text{ or } x-3=8 && \pmod{12} \\ \Leftrightarrow\quad & x=3 \text{ or } x=7 \text{ or } x=11 && \pmod{12}. \end{aligned}$$

"Or I could have observed that 3 is clearly one solution of the equation $3x = 9$ (mod 12). Furthermore, every four moves of 3 on a 12-hour clock returns me to where I started. Therefore, since $3 \cdot 3 = 9$, it follows that $3 \cdot (3 + 4) = 9$ and $3 \cdot (3 + 8) = 9$ (mod 12). That is, the other solutions of the equation are 7 and 11." *

"Okay, so if I *think* there are solutions, I don't have to plug in numbers. But how can I tell from just looking at an equation whether I should think it has solutions or not? I don't want to think it does and then go through the bother of trying to solve it when I thought wrong. For example, I know that an equation like $6x =$ (whatever) (mod 25) has solutions since 6 is relatively prime to 25, and so it has a multiplicative inverse. But how can I tell just from looking at an equation like $6x =$ (whatever) (mod 26) whether I should even *try* to solve it?"

"The next exercise should help you answer that question."

Exercise 4.7

1. What are the mod 20 multiples of 6? Which of the following equations have solutions?
   a. $6x = 10$ (mod 20) b. $6x = 9$ (mod 20)
   c. $6x = 16$ (mod 20)
2. What are the mod 20 multiples of 15? Which of the following equations have solutions?
   a. $15x = 5$ (mod 20) b. $15x = 9$ (mod 20)
   c. $15x = 10$ (mod 20)

---

* We'll discuss yet another method of solving equations with multiple solutions, such as $9x = 3$ (mod 12), in Section 7, where we'll merely "reduce" the equation to another one over a smaller mod. Any guesses on how we'll reduce $9x = 3$ (mod 12)?

Answers, 4.7

1. {0, 2, 4, 6, 8, 10, 12, 14, 16, 18}; a, c
2. {0, 5, 10, 15}; a, c

3. What are the mod 20 multiples of 12? Which of the following equations have solutions?
   a. $12x = 3 \pmod{20}$ b. $12x = 4 \pmod{20}$
   c. $12x = 10 \pmod{20}$
4. What are the mod 24 multiples of 15? Which of the following equations have solutions?
   a. $15x = 9 \pmod{24}$ b. $15x = 10 \pmod{24}$
   c. $15x = 12 \pmod{24}$
5. What are the mod 24 multiples of 20? Which of the following equations have solutions?
   a. $20x = 8 \pmod{24}$ b. $20x = 10 \pmod{24}$
   c. $20x = 15 \pmod{24}$
6. What are the mod 24 multiples of 10? Which of the following equations have solutions?
   a. $10x = 8 \pmod{24}$ b. $10x = 14 \pmod{24}$
   c. $10x = 15 \pmod{24}$
7. What are the mod 21 multiples of 18? Which of the following equations have solutions?
   a. $18x = 8 \pmod{21}$ b. $18x = 10 \pmod{21}$
   c. $18x = 15 \pmod{21}$

"Okay, Ant, so what this means is that if you have some equation like $6x = 14 \pmod{26}$, then since 6 isn't relatively prime to 26, it doesn't have a multiplicative inverse. So, you need to check whether the answer 14 is a multiple of 6 in $\mathbf{Z}_{26}$. And it will be if it's a regular multiple of 2, which it is, so the equation does."

"Come again?"

"It's simple. Take the gcd of 6 and 26, and get 2. Then note that 14 is a multiple of 2. So $6x = 14 \pmod{26}$ has solutions.

"Here's another example. Consider $20x = 12 \pmod{27}$. No problem since 20 has a multiplicative inverse. But change it to $21x = 12 \pmod{27}$ and you have to think: $\gcd(21,27) = 3$ and 3 is a factor of 12 so the equation has solutions. But the equation $21x = 14 \pmod{27}$ has no solutions since 3 isn't a factor of 14. I'm sure most people understand, so write all that up as a theorem."

Answers, 4.7 (cont.)

3. $\{0, 4, 8, 12, 16\}$; b
4. $\{0, 3, 6, 9, 12, 15, 18, 21\}$; a, c
5. $\{0, 4, 8, 12, 16, 20\}$; a
6. $\{0, 2, 4, \ldots, 16, 18, 20, 22\}$; a, b
7. $\{0, 3, 6, 9, 12, 15, 18\}$; c

Theorem 4.5

If a and b are elements of Z_n, $a \neq 0$, the equation $ax = b \pmod{n}$ has at least one solution if and only if $\gcd(a, n) \mid b$. When $\gcd(a, n) \mid b$, the equation has precisely $\gcd(a, n)$ solutions and they are spaced $\frac{n}{\gcd(a, n)}$ apart.

Exercise 4.8

Solve the following equations.

1. $4x = 10 \pmod{14}$
2. $12x = 10 \pmod{16}$
3. $27x = 15 \pmod{36}$
4. $8x = 3 \pmod{15}$
5. $6x = 5 \pmod{25}$
6. $9x = 12 \pmod{26}$
7. $15x = 10 \pmod{25}$
8. $18x = 12 \pmod{39}$
9. $18x = 20 \pmod{27}$

"But what do I do if you ask me to solve some equation like $13x = 11 \pmod{19}$? Just because I know that 13 has an inverse doesn't mean that I know which hat to pull it out of."

"There are a variety of techniques for finding multiplicative inverses, Gnam. I'll demonstrate some of them."

Answers, 4.8

1. 6, 13; $\gcd(4, 14) = 2$ and $2 \mid 10$. So the equation has solutions. $4x = 10 \Leftrightarrow 2(2x - 5) = 0 \Leftrightarrow 2x - 5 = 0$ or $2x - 5 = 7 \Leftrightarrow 2x = 5$ or $2x = 12 \pmod{14}$. $2x = 5 \pmod{14}$ has no solutions since $\gcd(2, 14)$ does not divide 5. But $2x = 12 \Leftrightarrow 2(x - 6) = 0 \Leftrightarrow x - 6 = 0$ or $x - 6 = 7 \Leftrightarrow x = 6$ or $x = 13$. (You also can reason as follows: If $4x = 10 \pmod{14}$ then $14 \mid (4x - 10)$, so $7 \mid (2x - 5)$. Hence, $r_7(2x - 5) = 0$ so $2r_7(x) = 5$. Thus $r_7(x) = 6$, where x is in $\mathbf{Z}_{14}$. Therefore $x = 6$ or $x = 13$.)
2. No solutions since $\gcd(12, 16) = 4$, and $4 \nmid 10$.
3. No solutions since $\gcd(27, 36) = 9$, and $9 \nmid 15$.
4. 6; $\gcd(8, 15) = 1$, so 8 has a mod 15 multiplicative inverse. $/8 = 2 \pmod{15}$. So, $8x = 3 \Leftrightarrow 2(8x) = 2(3) \Leftrightarrow x = 6$.
5. 5; $/6 = -4 \pmod{25}$, since $6(-4) = -24 = 1 \pmod{25}$. Hence, $6x = 5 \Leftrightarrow -4(6x) = -4(5) \Leftrightarrow x = -20 = 5 \pmod{25}$.
6. $\gcd(9, 26) = 1$ so 9 has a mod 26 multiplicative inverse. $/9 = 3 \pmod{26}$. Hence, $9x = 12 \Leftrightarrow 3(9x) = 3(12) \Leftrightarrow x = 10 \pmod{26}$.
7. $\gcd(15, 25) = 5$ and $5 \mid 10$, so the equation $15x = 10 \pmod{25}$ has solutions. They are 4, 9, 14, 19 and 24.
8. $\gcd(18, 39) = 3$ and $3 \mid 12$, so the equation $18x = 12 \pmod{39}$ has solutions. They are 5, 18, and 31.
9. $\gcd(18, 27) = 9$, and $9 \nmid 20$. So the equation $18x = 20 \pmod{27}$ has no solutions.

Exercise 4.9

Enter the mod 19 multiplicative inverses of 1, 2, 4, 5, 10, and 18 in the table below. These are all fairly simple to find. [Recall that $18 = -1 \pmod{19}$].

a	$/a \pmod{19}$	a	$/a \pmod{19}$
1		10	
2		11	
3		12	
4		13	
5		14	
6		15	
7		16	
8		17	
9		18	

"Please note the variety of strategies that I use to find the mod 19 multiplicative inverses of the other nonzero elements of $\mathbf{Z}_{19}$. (You may spot more clever approaches than I.) All of the following arithmetic is mod 19 arithmetic.

/3: Note that $3 \cdot 6 = -1$, and hence $3 \cdot (-6) = 1$. Thus, $/3 = -6 = 13$. Also, **/13** $= 3$.

/6: Again, $3 \cdot 6 = -1$, so $(-3) \cdot 6 = 1$. Hence, $/6 = -3 = 16$ and **/16** $= 6$.

/7: $7 = -12 = (-2) \cdot 6$. Now, $2 \cdot 10 = 1$, so $(-2)(-10) = 1$. We just showed that $/6 = 16$. Thus, $[(-2) \cdot 6] \cdot [(-10) \cdot 16] = 1$. So, $/7 = (-10) \cdot 16 = 9(16) = 9 \cdot 2 \cdot 8 = -1(8) = -8 = 11$. Also, **/11** $= 7$.

/8: $8 = 4 \cdot 2$. So, since $/2 = 10$ and $/4 = 5$ it follows that $8 \cdot (10 \cdot 5) = 1$. Thus $/8 = 10 \cdot 5 = 12$ and **/12** $= 8$. (Alternately: $8 \cdot 3 = 5$, so $8 \cdot (3 \cdot 4) = 5 \cdot 4 = 1$. Hence $/8 = 3 \cdot 4 = 12$.)

/9: $9 = -10$ and $(-10)(-2) = 1$. So $/9 = -2 = 17$. Also, **/17** $= 9$.

/14: $14 = -5$ and $/5 = 4$. Hence $/14 = -4 = 15$. Also, **/15** $= 14$."

Answers, 4.9:
$/1 = 1$; $/2 = 10$; $/4 = 5$; $/5 = 4$; $/10 = 2$; and $/18 = 18 \pmod{19}$

Exercise 4.10

Solve each of the following equations over $\mathbf{Z}_{19}$.

1. $11x = 5$
2. $5(x-2) = 14(1-x)$
3. $4(x+2) - 5(x-3) = 2x$
4. $12x + 15 = 5$
5. $3 - 8x = 11(x+2)$
6. $7(x-2) = 2/3$
7. $5x/9 = 2 - 3x/5$
8. $3/(x-1) = 2/(2x+5)$
9. $4(2x - 1/3) = x/5$

Exercise 4.11

1. Find the mod 23 multiplicative inverse of each nonzero element of $\mathbf{Z}_{23}$.
2. Solve over $\mathbf{Z}_{23}$: $5(x - 1/2) = 8$; $x/3 - 5 = 2x$; $9(x - 2) = 7$.

Supplementary Problem 4.5

Suppose that a and b have mod n multiplicative inverses. Prove the following.

1. $/(/a) = a \pmod{n}$
2. $/(ab) = (/a)(/b) \pmod{n}$
3. $/(-a) = -/a \pmod{n}$
4. If $ax = ay$ then $x = y \pmod{n}$.

Supplementary Problem 4.6

Show that b is a unit of an integral domain if and only if b has a multiplicative inverse.

Answers, 4.10

1. $11x = 5 \Leftrightarrow 7(11x) = 7(5) \Leftrightarrow x = -3 = 16 \pmod{19}$.
2. $5x - 10 = 14 - 14x \Leftrightarrow 0 = 5 \pmod{19}$. (Sorry, Gnam. No solutions.)
3. $4 = 3x \Leftrightarrow 13(4) = (13)(3x) \Leftrightarrow x = 14 \pmod{19}$.
4. $12x = -10 \Leftrightarrow 8(12x) = (8 \cdot -2)\,5 \Leftrightarrow x = 3 \cdot 5 = 15 \pmod{19}$.
5. $3 - 8x = 11x + 3 \Leftrightarrow 0 = 0 \pmod{19}$. All elements of $\mathbf{Z}_{19}$ are solutions.
6. $7(x-2) = 2/3 \Leftrightarrow 3(7x - 14) = 2 \Leftrightarrow 2x - 4 = 2 \Leftrightarrow 2x = 6 \Leftrightarrow x = 3$.
7. $5x/9 = 2 - 3x/5 \Leftrightarrow 5(5x) = 2{\cdot}5{\cdot}9 - 9(3x) \Leftrightarrow 6x = -5 - 8x \Leftrightarrow 14x = 14 \Leftrightarrow$ $x = 1 \pmod{19}$.
8. $x = 10$.
9. $x = 13$.

Answers, 4.11

1. $/1 = 1$; $/2 = 12$; $/3 = 8$; $/4 = 6$; $/5 = 14$; $/6 = 4$; $/7 = 10$; $/8 = 3$; $/9 = 18$; $/10 = 7$; $/11 = 21$; $/12 = 2$; $/13 = 16$; $/14 = 5$; $/15 = 20$; $/16 = 13$; $/17 = 19$; $/18 = 9$; $/19 = 17$; $/20 = 15$; $/21 = 11$; $/22 = 22$. Once you have figured out $/1$ through $/11$, $/22$ to $/12$ are just their additive inverses, respectively.
2. $x = 9$; $x = 20$; $x = 13$.

Supplementary Problem 4.7

> The cancellation property of multiplication states that if $ab = ac$ and $a \neq 0$ then $b = c$. Prove that the cancellation property holds for a commutative ring if and only if the ring has no zero divisors (that is, the ring is an integral domain).

"But what do I do if you give me an equation over some mod like 37 where I don't have a handy, dandy list of multiplicative inverses to refer to and none of the strategies you demonstrated seem to help? For example, suppose I had to find the mod 37 multiplicative inverse of 15?"

"I have a fall-back technique that I use whenever my mind goes blank. I just systematically start listing the multiples of 15 until it occurs to me how to combine two of them to obtain 1 or –1. Let me demonstrate:

$$1.\quad 1(15) = 15 = -22 \pmod{37}$$
$$2.\quad 2(15) = 30 = -7 \pmod{37}$$

"Aha! Look at the first and second equations:

$$1.\quad 1(15) = 15 \pmod{37}$$
$$2.\quad 2(15) = -7 \pmod{37}$$

"Multiply the second equation by 2, obtaining

$$2'.\quad 4(15) = -14 \pmod{37}$$

"Now, add	1.	$1(15) = 15 \pmod{37}$
to	2′.	$4(15) = -14 \pmod{37}$
to obtain:	3.	$5(15) = 1. \pmod{37}$

So $$/15 = 5 \pmod{37}.$$

Check: $5(15) = 5 \cdot 5 \cdot 3 = 25 \cdot 3 = -12 \cdot 3 = -36 = 1 \pmod{37}$.

"Here's another example: Find $/19 \pmod{61}$.

$$1.\quad 1(19) = 19 = -42 \pmod{61}$$
$$2.\quad 2(19) = 38 = -23 \pmod{61}$$
$$3.\quad 3(19) = 57 = -4 \pmod{61}$$

"Look at the first and third equations:

$$1.\quad 1(19) = 19 \pmod{61}$$

and $$3.\quad 3(19) = -4 \pmod{61}.$$

"Multiply equation 3 by 5 and add it to equation 1.
Equation 3 becomes:

$$3'.\quad 5(3)(19) = 5(-4), \text{ that is } (15)(19) = -20 \pmod{61}$$

So we have:	1. $1(19) = 19$	(mod 61)
and	3′. $15(19) = -20$	(mod 61)
Adding, we get	$16(19) = -1$	(mod 61)
Hence	$/19 = -16 = 45$	(mod 61).

Check: $19(45) = 19(3)(15) = 57(15) = (-4)(15) = -60 = 1 \pmod{61}$.

"This time, I listed the multiples of 19 until I found a linear combination of two equations that produced –1, rather than 1."

Exercise 4.12

1. Find the mod 61 multiplicative inverses of 18, 8, 29, and 35.
2. Solve over $\mathbf{Z}_{61}$: $3(6x - 2) = 57$; $2(3x - 1) = 59x + 3$; $x/8 + 3x/32 = 5$.
3. Solve $3(2x + 1) - 7(2 - 3x) = 46 \pmod{71}$; $29x = 32 \pmod{71}$.

"It's quite a coincidence, Ant, that you can solve equations like $9x - 18 = 7 \pmod{61}$ pretty much like you do in ordinary algebra."

"It's no coincidence, Gnam. The modular systems are all commutative rings. And in *any* commutative ring the equation $ax + b = c$ has a unique solution if a has a multiplicative inverse. You simply argue:"

Answers, 4.12

1. $/18 = /(6 \cdot 3) = (-10)(-20) = 17$; $/8 = /(4 \cdot 2) = (-15)(-30) = 23$; $2(29) = 58 = -3$. So $(20 \cdot 2)(29) = (20)(-3) = 1$. $/29 = 40$; $1(35) = 35$ and $2(35) = 9$. Subtract the first equation from 4 times the second one. $/35 = 7$.
2. 34; 54; 49, since $x/8 + 3x/32 = 5 \Leftrightarrow 32[x/8] + 32[3x/32] = 32 \cdot 5 \Leftrightarrow [x/8][8 \cdot 4] + 3x = 4 \cdot 32 + 1 \cdot 32 \Leftrightarrow 4x + 3x = 2(2 \cdot 32) + 32 \Leftrightarrow 7x = 2(3) + 32 = 38 \Leftrightarrow x = 35(38) = 19(2 \cdot 35) = 19(9) = 3(3 \cdot 19) = 3(57) = 3(-4) = -12 = 49 \pmod{61}$.
3. $3(2x + 1) - 7(2 - 3x) = 46 \Leftrightarrow 27x = 57 \Leftrightarrow 9x = 19 \Leftrightarrow x = 8(19) = 10 \pmod{71}$; $x = 6$.

Suppose that a has a multiplicative inverse. Then,

$ax + b = c$

$\Leftrightarrow (ax + b) + (-b) = c + (-b)$ [existence of additive inverses]

$\Leftrightarrow ax + (b + [-b]) = c + (-b)$ [associative property of addition]

$\Leftrightarrow ax + 0 = c + (-b)$ [definition of additive inverses]

$\Leftrightarrow ax = c + (-b)$ [identity property of 0]

$\Leftrightarrow /a(ax) = /a \cdot (c + [-b])$ [existence of multiplicative inverse of a]

$\Leftrightarrow (/a \cdot a)x = /a \cdot (c + [-b])$ [associative prop. of multiplication]

$\Leftrightarrow 1x = /a \cdot (c + [-b])$ [definition of multiplicative inverses]

$\Leftrightarrow x = /a \cdot (c + [-b])$ [identity property of 1].

"Note that not once did I use commutivity! That's because, compared to associativity, commutivity is a second-class citizen. Look at the ring of matrices, for example. Matrix multiplication is associative, but not commutative, and what would linear algebra be without matrices? And what's the most important operation on functions? Composition! It's associative, but not commutative. Algebraists commonly deal with noncommutative operations…"

"Help! Halt! Enough! It's time for another exercise!"

Exercise 4.13

Assume that p is a prime number greater than 8.

1. Show that $2/3 + 4/3 = 2 \pmod{p}$.
2. Show that $(8)(3/4) = 6 \pmod{p}$.
3. Show that $(2/3)(3/4) = 1/2 \pmod{p}$.

"I meant that it was time for an algebra problem, Ant."

"Very well, let's solve two equations in two unknowns. The system

$$4x + 3y = 4 \pmod{19}$$
$$7x - 2y = 2 \pmod{19}$$

is equivalent to the system

$$7x - 9y = -12 \pmod{19}$$
$$7x - 2y = 2 \pmod{19}.$$

Answers, 4.13

1. $2/3 + 4/3 = (2 + 4)/3 = 6/3 = 2(3)/3 = 2(1) = 2 \pmod{p}$.
2. $8(3/4) = 8 \cdot /4 \cdot 3 = 2 \cdot 4 \cdot /4 \cdot 3 = 2 \cdot 1 \cdot 3 = 6 \pmod{p}$.
3. $(2/3)(3/4) = 2{\cdot}3 \cdot /3 \cdot /4 = 2/4 = 2/(2{\cdot}2) = 2\,(/2 \cdot /2) = /2 = 1 \cdot /2 = 1/2 \pmod{p}$.

I got this by multiplying the first equation by –3. Subtracting the first equation above from the second one, we obtain:

$$7y = 14 \pmod{19}$$

$$y = 2 \pmod{19}.$$

Substituting 2 for y in either of the original equations yields $x = 9$. Thus the solution of this system is (9, 2)."

Exercise 4.14

Find the simultaneous solutions to each of the following pairs of linear equations over the indicated mod.

1. $2x + 5y = 11$ and $5x - 2y = 13 \pmod{19}$
2. $4x - 2y = -2$ and $3x - 5y = 7 \pmod{13}$
3. $6x + 2y = 12$ and $7x + 3y = 1 \pmod{17}$
4. $x + y = 0$ and $x - y = 2 \pmod{10}$

"That about does it for ninth-grade algebra, Ant. I guess we can roll up this section now."

"Once again you have forgotten quadratics, Gnam. Given any quadratic equation $ax^2 + bx + c = 0 \pmod{n}$, we can apply the quadratic formula provided that $2a$ has a mod n multiplicative inverse."

Quadratic Formula

If $2a$ has a mod n multiplicative inverse, then x_0 is a solution of $ax^2 + bx + c = 0 \pmod{n}$ if and only if $x_0 = (-b + k)/(2a)$ where k is a mod n square root of $b^2 - 4ac$.*

Example: Find all x in $\mathbf{Z}_9$ such that $2x^2 - x + 3 = 0 \pmod{9}$.

Solution: $/(2a) = /4 = 7 \pmod{9}$ and $b^2 - 4ac = (-1)^2 - 4(2)(3) = 4 \pmod{9}$.

The mod 9 square roots of 4 are 2 and 7. So, the solutions of the quadratic equation $2x^2 - x + 3 = 0 \pmod{9}$ are: $x_0 = [-(-1) + 2] \cdot 7 = 3$ and $x_0 = [-(-1) + 7] \cdot 7 = 2 \pmod{9}$.

Answers, 4.14

1. (3, 1)
2. (2, 5)
3. (0, 6)
4. There are *two* solutions: (6, 4) and (1, 9). When solving linear equations over composite mods, take care to obtain all the solutions.

* If $n \leq 4$, substitute $r_n(4)$ for 4 in the formula.

"Incidentally, $2x^2 - x + 3 = 2(x-3)(x-2) \pmod 9$."

"And once again you've forgotten that I don't like to solve quadratics by plugging numbers into formulas. I prefer to complete the square."

"The formula is *derived* by completing the square in the general case, Gnam. But you can find the solutions using any procedure you like."

"Fine. This is how I would solve $2x^2 - x + 3 = 0 \pmod 9$."

$$2x^2 - x + 3 = 0 \qquad \pmod 9$$

$$\Leftrightarrow \qquad x^2 - 5x + 6 = 0 \qquad \pmod 9$$

$$\Leftrightarrow \qquad x^2 + 4x + (6-2) + 2 = 0 \qquad \pmod 9$$

$$\Leftrightarrow \qquad (x+2)^2 = -2 = 7 \qquad \pmod 9$$

$$\Leftrightarrow \qquad x + 2 = 4 \text{ or } x + 2 = 5 \qquad \pmod 9$$

(4 and 5 are the mod 9 square roots of 7)

$$\Leftrightarrow \qquad x = 2 \text{ or } x = 3 \qquad \pmod 9.$$

"Well done, Gnam."

"Not really, Ant. I just realized that I went through much too much bother. Once I had $x^2 - 5x + 6 = 0$, I could have just factored it into $(x-3)(x-2) = 0$ and gotten the solutions that way!"

"You need to take care over composite mods, Gnam. Clearly 3 and 2 are both solutions of $(x-3)(x-2) = 0 \pmod 9$. But 9 is a composite, so $\mathbf{Z}_9$ has zero divisors. From $(x-3)(x-2) = 0$ (mod some composite), you can't deduce that $x - 3 = 0$ or $x - 2 = 0$. You could miss solutions! For example, both 2 and 3 are solutions of $(x-3)(x-2) = 0 \pmod{10}$, but so are 7 and 8 since $(7-3)(7-2) = 0 \pmod{10}$ and $(8-3)(8-2) = 0 \pmod{10}$. The equation $(x-3)(x-2) = 0 \pmod{10}$ has four solutions! Indeed, one can't even apply the quadratic formula to $x^2 - 5x + 6 = 0 \pmod{10}$ since 2 has no mod 10 multiplicative inverse. And hence, completing the square isn't viable either."

"So how am I supposed to solve that equation, Ant?"

"By plugging in all the numbers, Gnam. What else?"*

---

*Sorry, but I couldn't resist saying that to Gnam. There do exist techniques for finding the solutions of quadratics over composite mods which don't involve plugging in numbers, but they are beyond the scope of this book.

Exercise 4.15

Find all the solutions of each of the following equations over the indicated mod.

1. $2x^2 + x + 2 = 0 \pmod 5$
2. $6 + 5x = 6x^2 \pmod 7$
3. $x^2 + 5x = 0 \pmod 6$
4. $x^2 - 8x = 0 \pmod{15}$
5. $x^2 - 2x + 2 = 0 \pmod{13}$
6. $\dfrac{2}{x+3} + \dfrac{5}{x-1} = 6 \pmod{17}$
7. $(4x - 3)(x + 4) = (1 - 3x)(3x + 2) \pmod 9$
8. $(2x + 1)(3x - 4) = (4 - x)(2 + x) + 11 \pmod{19}$

"It's really strange, Ant, that an element can have more than two square roots. For example, 4 has four different mod 15 square roots."

"That can happen over a composite mod, Gnam, but not over a prime mod. In any mod, if $k = a^2$ and $k = b^2$, then $a^2 = b^2$ and so $a^2 - b^2 = 0 \pmod n$. Thus, $(a - b)(a + b) = 0 \pmod n$. If n is composite, it doesn't follow that either $a - b$ or $a + b$ has to be zero. But if n is prime, $\mathbf{Z}_n$ has no zero divisors, and therefore $a - b = 0$ or $a + b = 0 \pmod n$. Thus, $a = b$ or $a = -b \pmod n$. So an element k of a prime mod has at most two square roots.

Answers, 4.15

1. $\{1\}$
2. $\{4, 5\}$
3. The quadratic formula can't be used here because 2 has no mod 6 multiplicative inverse. The solutions of this equation are 0, 1, 3, and 4. $x(x + 5) = 0 \pmod 6$ does not imply that one of the factors must be 0, since Z_6 has mod 6 zero divisors.
4. The quadratic formula can be used here since $/(2a) = /2 = 8 \pmod{15}$. $b^2 - 4ac = 7^2 - 4(1)(0) = 4 \pmod{15}$. Now, 4 has four mod 15 square roots, namely 2, 7, 8, and 13. (I made up a table of mod 15 squares to find these.) Substituting each of these into the quadratic formula, we obtain the four solutions of the equation $x^2 - 8x = 0$, namely 0, 3, 5, and 8. By the way, there are two different ways of factoring $x^2 - 8x$ over Z_{15}: $x(x-8)$ and $(x - 3)(x - 5)$.
5. Using the quadratic formula, the solutions are $(2 + 3)/2 = 9$ and $(2 + 10)/2 = 6$. Note that $x^2 - 2x + 2 = (x - 9)(x - 6) \pmod{13}$.
6. $\{4, 15\}$
7. No solutions
8. $\{3, 17\}$

"Elements of modular systems which are perfect squares are called **quadratic residues**. In particular, for each n, 0 and 1 are mod n quadratic residues since $0^2 = 0$ and $1^2 = 1$. Making a table of perfect squares is a tedious way of deciding whether a particular number is a perfect square. There is a more elegant technique for doing this which uses the Quadratic Reciprocity Theorem, one of the prettiest theorems in mathematics."

"You can reciprocate quadratically some other time, Ant. My brain is starting to hurt. How about asking me something easy."

Exercise 4.16

Despite whatever evidence you may have to the contrary, it is now 11:00 p.m., Wednesday, Nov. 3.

1. What time will it be in 243 hours?
2. What time was it 243 hours ago?
3. What day of the week will it be in 1475 days?
4. What day of the week was it 1475 days ago?
5. What month will it be in 125 months?
6. What month was it 125 months ago?

Supplementary Problem 4.8

Suppose that p is a prime. Prove that if $x = /x \pmod p$ then $x = 1$ or $x = p - 1$. (In other words, 1 and –1 are the only elements of $\mathbf{Z}_p$ which are their own mod p multiplicative inverses.)

Supplementary Problem 4.9

Prove that if p is prime, then $\mathbf{Z}_p$ has precisely $1 + (p-1)/2$ quadratic residues.

Supplementary Problem 4.10

1. What is 10! (mod 11)? 2. What is 16! (mod 17)?
3. Prove that if p is prime, then $(p-1)! = -1 \pmod p$. (This is known as **Wilson's Theorem.**)

Answers, 4.16

1. 2:00 A.M. $r_{24}(243) = 3$ and $11 + 3 = 2 \pmod{12}$.
2. 8:00 P.M. $11 - 3 = 8 \pmod{12}$.
3. Monday. $r(1475) = 5 \pmod 7$ and 5 days from Wednesday is the same weekday as two days ago, which is Monday.
4. 2 days from now is Friday.
5. November is the 11th month, so the answer is the 4th month, April.
6. $11 - 5 = 6$. The 6th month is June.

Supplementary Problem 4.11

Derive the quadratic formula.

Supplementary Problem 4.12

1. There are 18 quadratic equations of the form $ax^2 + bx + c = 0$ $(a \neq 0)$ over $\mathbf{Z}_3$. However, any such equation with leading coefficient 2 is equivalent to one with leading coefficient 1. For example, the equation $2x^2 + x + 1 = 0$ (mod 3) is equivalent to the equation $x^2 + 2x + 2 = 0$ (mod 3) since $2x^2 + x + 1 = 2(x^2 + 2x + 2)$ (mod 3). List the nine distinct quadratic equations over $\mathbf{Z}_3$.
2. Either by factoring or by using the quadratic formula, solve each of these 9 quadratic equations over $\mathbf{Z}_3$, where possible. (Some of the equations will have no solutions in $\mathbf{Z}_3$.)

"The equations $x^2 + bx + c = 0$ (mod 3) which have no solutions in $\mathbf{Z}_3$ are those for which the discriminant, $b^2 - 4c$, is 2. This is because 2 has no mod 3 square roots, whereas both 0 and 1 do. Since $\mathbf{Z}_3$ is not large enough to allow us to solve all quadratic equations over $\mathbf{Z}_3$, let us 'fix' the situation by enlarging our set (much as we do when we extend the real number system to the complex number system in order to ensure that polynomials will have roots). We need to add a square root of 2. But note that $2 = -1$ (mod 3). A common name for a square root of -1 is i, so let's toss i into our set. Of course, we want our extended number system to have all the nice properties of $\mathbf{Z}_3$. For, example, it must be closed under addition and multiplication. So, if we are going to toss i into the set, we also have to toss in $i + i = 2i$ and $2 + i$ and so on.

"Define $\mathbf{Z}_3(i) = \{r + si\text{: } r, s \text{ are elements of } \mathbf{Z}_3\}$. Thus, $\mathbf{Z}_3(i)$ not only contains the three elements of $\mathbf{Z}_3$ (where $s = 0$), but six other elements (where $s \neq 0$). Some examples of computation in $\mathbf{Z}_3(i)$ are $(2 + i) + (2 + 2i) = 1$ (mod 3), $(1 + i) + (2 + i) = 2i$ (mod 3), $(2 + i)(2 + 2i) = 2$ (mod 3) and $2(1 + i) = 2 + 2i$ (mod 3)."

3. Make up a 9 by 9 addition chart and a 9 by 9 multiplication chart for $\mathbf{Z}_3(i)$, showing the mod 3 sum and product of every pair of elements of $\mathbf{Z}_3(i)$.

4. Make a table which lists the additive inverse of each element of $\mathbf{Z}_3(i)$. Make a table which lists the multiplicative inverse of each nonzero element of $\mathbf{Z}_3(i)$.
5. Now solve the rest of the 9 quadratic equations you listed in part 1 above over $\mathbf{Z}_3(i)$.*

"The essential structural difference between the prime modular systems and the composite modular systems is not merely that the former have no zero divisors (that is, are integral domains) whereas the latter do, but, more importantly, that each nonzero element in a prime modular system has a multiplicative inverse. Integral domains in which all elements but 0 have multiplicative inverses are called **fields**.

"Examples of fields include the rational numbers, the real numbers, the complex numbers and, for prime p, $\mathbf{Z}_p$ under $+_p$ and $\cdot_p$. ($\mathbf{Z}_3(i)$ under $+_3$ and $\cdot_3$, defined in Supplementary Problem 4.12, is also a field.) Although all fields are integral domains, the converse isn't true. For example, the integers are an integral domain, but not a field.

"Many of the proofs used to establish properties about $\mathbf{Z}_{11}$, such as those given in Supplementary Problem 2.6, can easily extend to any field. There is great power in doing this! You never can tell when a new algebraic structure might come to your attention, and if you are able to classify it as an integral domain or a field or whatever, you already know a lot about the structure. By the way, Gnam, where are you? I haven't seen you in a while."

"You want to talk about fields? You want to discuss problems? Okay, how about this. You have avoided the Black Magic Device Problem for so long that it seems that the device is outside your domain of expertise and is not a field for discussion."

"Nonsense! We will analyze it in the next section."

"Then let's begin the next section now!"

"First I want to visit a rather special place. After that, the next section can begin."

---

*Actually, we aren't "fixing" things all that well. While all quadratics with coefficients from $\mathbf{Z}_3$ have roots in $\mathbf{Z}_3(i)$, there are quadratics with coefficients from $\mathbf{Z}_3(i)$ which don't have roots in $\mathbf{Z}_3(i)$. On the other hand, when we extend the real numbers to the complex numbers, not only do all polynomials (of degree greater than 0) with real coefficients have complex roots, but so do all polynomials with complex coefficients.

Section Five

Remainder Tests

Dear Readers,

Just before this section began, Ant took me to a remarkable place where a row of machines (all looking like the one I've drawn below) seemed to stretch into eternity. Ant told me that for each n greater than 1, there is a remainder machine, r_n, which works as follows: When you feed the r_n machine the integer k, the machine spews out $r_n(k)$, and does so *without going through the long division process.* Among these machines was the r_{13} machine which does the same thing as the Black Magic Device, only it works for all integers. Ant solemnly promised me that in this section we would learn how some of these machines work. However, Ant has promised that before....

Sincerely,
Gnam

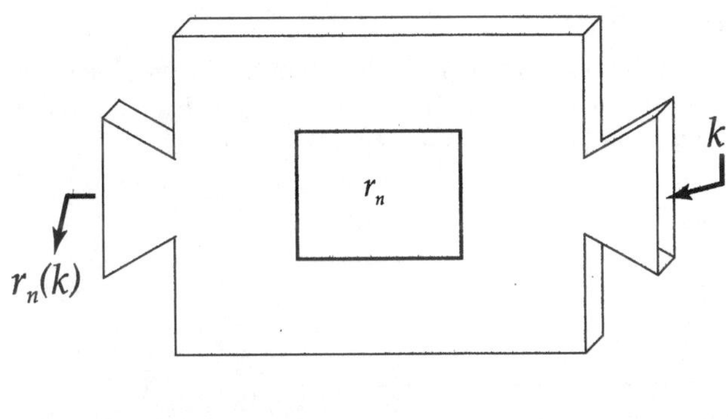

"Gnam, as I promised you...and we all know my word is good...we are going to develop remainder tests in this section. These tests will make repeated use of the Super-Important Theorem, which possibly you recall."

Super-Important Theorem

If a, b, and n are integers and $n > 1$, then

$$r_n(a+b) = r_n(a) +_n r_n(b) \quad \text{and}$$

$$r_n(a \cdot b) = r_n(a) \cdot_n r_n(b).$$

Exercise 5.1

Without using the long division process, find $r_5(k)$ for each of the following k: 26; 39; 6823497; –3487987.

"Ant, you don't need the Super-Important Theorem to do the last exercise. Any fifth grader can tell you remainders when you divide by 5. Just look at the ones digit."

"But the Super-Important Theorem explains *why* you need only examine the ones digit. For example, consider $r_5(6823497)$:

$$6823497 = 682349 \cdot 10 + 7.$$

Hence, by The Super-Important Theorem:

$$r_5(6823497) = r_5(682349) \cdot_5 r_5(10) +_5 r_5(7)$$

but $$r_5(10) = 0,$$

and thus $r_5(682349) \cdot_5 r_5(10) = 0$.

Therefore $$r_5(6823497) = r_5(7) = 2.$$

"In general, any integer can be expressed as $k \cdot 10 + j$, and $r_5(k \cdot 10 + j) = r_5(j)$ by the Super-Important Theorem."

Exercise 5.2

Are there any machines other than the r_5 machine which need consider only the ones digit? List all such machines. How do you know your list is complete?

"The obvious question to ask next is:"

Answers, 5.1
1; 4; 2; 3.

Answers, 5.2
Yes, r_2 and r_{10}. These are the only ones, because the only factors of 10 which are bigger than 1 are 2, 5 and 10.

Exercise 5.3

1. Which machines need consider precisely the rightmost two digits (that is the ones digit and the tens digit)? List all such machines. For each machine r_n which qualifies, factor n into a product of primes.
2. Which machines need consider precisely the rightmost three digits? For each such machine r_n factor n into a product of primes.
3. What is the next question? What is its answer?
4. Generalize.

Exercise 5.4

Find the following remainders without going through the long division process.

1. $r_8(2936641)$
2. $r_{40}(6823416)$
3. $r_{25}(889345234)$
4. $r_{32}(-15483964331)$

Supplementary Problem 5.1 (For those familiar with number bases)

The results of the previous two exercises are dependent upon the fact that we use a base ten system of numeration. Suppose that we used base twelve instead. Then for which n is the remainder mod n of the positive integer k equal to the remainder mod n of the rightmost t digits of k? Answer the same question if we used a base thirty system of numeration.

"We are now going to consider machines which, like r_{11}, must examine *all* of the digits of the integer which is input. Let's start with r_3. Observe that by the Super-Important Theorem

---

Answers, 5.3

1. Any positive integer can be expressed as $100k + j$, and $r_n(100k + j) = r_n(100) \cdot_n r(k) +_n r_n(j)$. We are seeking those n such $r_n(100) = 0$, but $r_n(10) \neq 0$. Thus n must be a factor of 100 but not of 10. The n that work are: $4 = 2^2$; $20 = 2^2 \cdot 5$; $25 = 5^2$; $50 = 2 \cdot 5^2$; and $100 = 2^2 \cdot 5^2$.
2. 2^3; $2^3 \cdot 5$; $2^3 \cdot 5^2$; $2^3 \cdot 5^3$; $2 \cdot 5^3$; $2^2 \cdot 5^3$; and 5^3.
3. Those r_n where n factors into $2^r \cdot 5^s$ and at least one of the exponents r and s is 4, and the other exponent is between 0 and 4, inclusive.
4. The machine r_n needs to consider precisely the rightmost t digits if and only if n factors into $2^r \cdot 5^s$ where at least one of the exponents is t and the other exponent is between 0 and t, inclusive.

Answers, 5.4

1. $r_8(641) = 1$
2. $r_{40}(416) = 16$
3. $r_{25}(34) = 9$
4. $r_{32}(64331) = 11$ so $r_{32}(-64331) = 21$

$$\begin{aligned} r_3(10) &= 1 \\ r(100) &= r(10)\cdot r(10) = 1\cdot 1 = 1 \pmod 3 \\ r(1000) &= r(10)\cdot r(100) = 1\cdot 1 = 1 \pmod 3 \end{aligned}$$

and, in general, for any positive integer s,

$$r_3(10^s) = [r_3(10)]^s = 1^s = 1 \pmod 3.$$

"Now we shall figure out what $r_3(746)$ is:

$$746 = 7(100) + 4(10) + 6.$$

So, by the Super-Important Theorem:

$$\begin{aligned} r(746) &= r(7)r(100) + r(4)r(10) + r(6) && \pmod 3 \\ &= 1\cdot 1 + 1\cdot 1 + 0 && \pmod 3 \\ &= 2 && \pmod 3. \end{aligned}$$

Similarly,

$$\begin{aligned} r(6482) &= r(6)r(1000) + r(4)r(100) + r(8)r(10) + r(2) && \pmod 3 \\ &= 0\cdot 1 + 1\cdot 1 + 2\cdot 1 + 2 && \pmod 3 \\ &= 2 && \pmod 3 \end{aligned}$$

and

$$\begin{aligned} r(2513028) &= r(2) + r(5) + r(1) + r(3) + r(0) + r(2) + r(8) && \pmod 3 \\ &= 2 + 2 + 1 + 0 + 0 + 2 + 2 && \pmod 3 \\ &= 0 && \pmod 3. \end{aligned}$$

"In general, if k is a positive integer, $r_3(k)$ is just the mod 3 sum of the digits of k."*

Exercise 5.5

Find $r_3(k)$ for each of the following k.

1. 68747
2. 120978
3. 123123123
4. –45667
5. –4291
6. $326 \cdot 15422 + 976312 \cdot 54872$

"Note that it is also the case that $r_9(10) = 1$, and so for each positive integer s, $r_9(10^s) = [r_9(10)]^s = 1^s = 1 \pmod 9$. Hence, to

---

*What I really mean is "$r_3(k)$ is the mod 3 sum of the mod 3 remainders of the digits of k." It's just that I seem to confuse Gnam whenever I say it right.

Answers, 5.5

1. 2
2. 0
3. 0
4. 2
5. 2
6. $2(2) + 1(2) = 0 \pmod 3$

find the mod 9 remainder of the positive integer k, it suffices to add mod 9 the digits of k."

Exercise 5.6

Find $r_9(k)$ for each of the following k.

1. 134568
2. 123456789
3. –312
4. $432 \cdot 598 + 15$
5. $345 \cdot 24367543$

Exercise 5.7

For which n greater than 1, other than 3 and 9, can one find the mod n remainder of a positive integer k by adding mod n the digits of k?* How do you know your list is complete?

"When I was in grade school, calculators hadn't been invented. So we were taught a method of checking our computations called 'casting out nines'. Consider the following multiplication problem.

$$\begin{array}{r} 7682 \\ \times 4296 \\ \hline 32668482 \end{array}$$

"We add 7 + 6 + 8 + 2 getting 23, and then add 2 + 3, getting 5.

"We add 4 + 2 + 9 + 6 getting 21 and then add 2 + 1 getting 3.

"Now, $5 \cdot 3 = 15$. (We multiply since this is a multiplication problem.) And 1 + 5 = 6. So when we 'cast the nines' out of our answer, if the sum isn't 6, our answer is wrong.

"3 + 2 + 6 + 6 + 8 + 4 + 8 + 2 = 39 and 3 + 9 = 12 and 1 + 2 = 3. That means that our answer is wrong.

"Although none of my teachers ever explained why this works, I realize now that if $7682 \cdot 4296 = 32668482$ then

---

Answers, 5.6

1. 0
2. 0
3. $-6 = 3 \pmod 9$
4. $0 + 6 = 6 \pmod 9$
5. $3(7) = 3 \pmod 9$

*Once again, Gnam gives me one of those quizzical looks when I say it right: "...by adding mod n the mod n remainders of the digits of k."

Answer, 5.7

There are none. The remainder tests for 3 and 9 are due to $r_3(10) = r_9(10) = 1$. There is no other n greater than 1 such that $r_n(10) = 1$.

$r(7682) \cdot r(4296) = r(32668482) \pmod 9$. We found the mod 9 remainders of each of these three numbers by adding their digits. This can be done in a less cumbersome way by adding mod 9 rather than adding in ordinary arithmetic. For example,

$$\begin{aligned} 7+6+8+2 &= (7+2)+(6+8) \pmod 9 \\ &= 0+5 \pmod 9 \\ &= 5 \pmod 9. \end{aligned}$$

"You can also check addition and subtraction problems by casting out nines. But there is a problem with this method: If the answer is wrong by a multiple of 9, then the check won't catch it!"

Exercise 5.8

"Check" the answers to the following problems by casting out nines.

1.
$$\begin{array}{r} 43762 \\ 24957 \\ +\ 6234 \\ \hline 74853 \end{array}$$

2.
$$\begin{array}{r} 618417 \\ -\ 395889 \\ \hline 222528 \end{array}$$

3.
$$\begin{array}{r} 31435 \\ \times\ 2231 \\ \hline 79131485 \end{array}$$

Supplementary Problem 5.2 (For those familiar with number bases)

The remainder tests for 3 and 9 described in the previous pages are dependent upon the fact that we use a base ten system of numeration. Suppose that we used a base nine system of numeration. Then for which n is $r_n(k)$ equal to the mod n sum of the digits of the positive integer k?

"And now we come to the Black Magic Device Problem. Recall that the Black Magic Device gives you the mod 13 remainder of any number between –1000 and 1000. The Black Magic Device Problem is to figure out a way to use the Black Magic Device to compute the remainder of *any* integer when it is divided by 13.

Answers, 5.8

1. $r(43762) + r(24957) + r(6234) = 4 + 0 + 6 = 1 \pmod 9$ but $r_9(74853) = 0$. So the answer is wrong.
2. $r(618417) - r(395889) = 0 - 6 = 3 \pmod 9$ and $r_9(222528) = 3$. The answer still could be wrong by a multiple of 9. (It isn't. The answer is correct.)
3. $r(31435) \cdot r(2231) = 7(8) = 2 \pmod 9$ and $r_9(79131485) = 2$. However, the answer is wrong by a multiple of 9. The check doesn't catch the error.

"Now, $r_{13}(10) = 10 = -3 \pmod{13}$. Both 10 and -3 are too far from 0 for my taste, so let's continue with $r_{13}(10^2) = (-3)(-3) = 9 = -4 \pmod{13}$. I'm still not happy. Let's go on to $r_{13}(10^3)$.

"This time we have $r_{13}(10^3) = r_{13}(10^2)\, r_{13}(10) = (-4)(-3) = 12 = -1 \pmod{13}$. Aha! I very much like computing with -1. Now, since $r_{13}(10^3) = -1$, it follows that $r_{13}(10^6) = 1$ and $r_{13}(10^9) = -1$ (mod 13), *etc*. In other words:

$$r(10^{3s}) = 1 \pmod{13} \quad \text{if } s \text{ is even, and}$$
$$r(10^{3s}) = -1 \pmod{13} \quad \text{if } s \text{ is odd.}$$

So: $r_{13}(296,684) = r_{13}(296 \cdot 10^3 + 684) = r_{13}(296) \cdot (-1) + r_{13}(684)$ (mod 13).

"If we offer the Black Magic Device the number 296, it will respond with 10. And if we offer the Black Magic Device the number 684, it will respond with 8. Thus:

$$\begin{aligned} r(296{,}684) &= r(296)\cdot(-1) + r(684) && \pmod{13} \\ &= 10(-1) + 8 && \pmod{13} \\ &= -2 && \pmod{13} \\ &= 11 && \pmod{13}. \end{aligned}$$

Similarly,

$$\begin{aligned} r(68{,}141{,}525) &= r(68)r(10^6) + r(141)r(10^3) + r(525) && \pmod{13} \\ &= r(68) - r(141) + r(525) && \pmod{13}. \end{aligned}$$

Now we use the Black Magic Device to learn that

$$r_{13}(68) = 3,\ r_{13}(141) = 11, \text{ and } r_{13}(525) = 5.$$

Thus

$$\begin{aligned} r(68{,}141{,}525) &= 3 - 11 + 5 && \pmod{13} \\ &= 10 && \pmod{13}." \end{aligned}$$

"But what about people who don't have access to the Black Magic Device, Ant? Do they have to divide 68, 141, and 525 by 13 to see what the remainders are?"

"Not if you feel, as I do, that dividing by 25 is easier than dividing by 13. Note that $r(25) = -1 \pmod{13}$. So, if the Black Magic Device isn't handy, you can argue as follows:

$$r_{13}(68, 141, 525) = r_{13}(68 - 141 + 525) = r_{13}(452).$$

"Now, there are four 25's in a hundred, and so there are eighteen of them in 452."

$$452 = (18)(25) + 2.$$

$$\begin{aligned} r(452) &= r(18)r(25) + r(2) \quad &(\text{mod } 13) \\ &= 5(-1) + 2 &(\text{mod } 13) \\ &= -5 + 2 &(\text{mod } 13) \\ &= 10 &(\text{mod } 13). \end{aligned}$$

"Similarly, $r(296{,}684) = r(-296 + 684) = r(388) = r(15 \cdot 25 + 13) = 2 \cdot (-1) + 0 = 11 \pmod{13}$."

Exercise 5.9

Without dividing any numbers by 13, compute the mod 13 remainders of the following.

1. 215, 678 2. 129, 476 3. −34, 898
4. 317, 114 5. 222, 333, 444 6. 159, 607, 200, 013

Exercise 5.10

Develop one or more remainder tests for each of the following divisors: 6; 7; 12; 15; 24; 30; 45; 90.

"In Sections 6 and 7 we will apply the theory of remainders to a variety of situations. The duration of this section will be devoted to an analysis of problems of the following type:

Supplementary Problem 5.3

Find $r_7(216310^{65374})$.

Answers, 5.9

1. $r(215{,}678) = r(678 - 215) = r(463) = r(18 \cdot 25 + 13) = -5 + 0 = 8 \pmod{13}$
2. $r(129{,}476) = r(476 - 129) = r(347) = r(13 \cdot 25 + 22) = 0 + 9 = 9 \pmod{13}$
3. $r(-34{,}898) = -r(898 - 34) = -r(864) = -r(34 \cdot 25 + 14) = -(-8 + 1) = -(-7) = 7 \pmod{13}$
4. 5 5. 8 6. 1

Answer, 5.10

I may have many talents, but I am not clairvoyant! Students of my imaginary friend, Andrea Rothbart (actually, she is a true friend, just an imaginary character), have created remainder tests for each of the above numbers (and many others). For example, in seeking a remainder test for 6, some of her students have noted that $r_6(25) = 1$ and then proceeded from there. Others have analyzed the situation by cleverly combining the outputs of the r_2 and r_3 machines. Still others have created a remainder test by first examining $r_6(10^k)$ for $k \geq 1$ and then developing a method which involves 'casting out threes'! Once you understand the basic strategies, you might create remainder tests that are new to Andrea and me.

"Do you have any idea how large a number 216310^{65374} is, Gnam? No computer could possibly calculate it. Nevertheless, human beings are endowed with a brain so powerful we can figure out what the remainder of this number is when it's divided by 7."

"I think I'll cast myself out of this section, Ant. See ya later!"

"No, don't go! There is a very powerful strategy applicable to many problems. If a problem looks difficult, try to solve a similar but simpler problem. You may see a technique that applies to the more difficult problem or a pattern that generalizes. In this case, let us begin by seeking mod 3 (rather than mod 7) remainders of powers of 1, then powers of 2, then powers of 3, *etc.* in the hope that we'll observe a pattern."

Simpler Problem

1. What is $r_3(1^{237564})$? $r_3(1^k)$ for any positive integer k?
2. What is $r_3(2)$? $r_3(2^2)$? $r_3(2^3)$? $r_3(2^4)$? $r_3(2^5)$? $r_3(2^{23561})$? $r_3(2^{244})$? Generalize.
3. What is $r_3(3^4)$? $r_3(3^{26874})$? $r_3(3^k)$ for any positive integer k?
4. What is $r_3(4)$? $r_3(4^2)$? $r_3(4^3)$? $r_3(4^5)$? Generalize.
5. What is $r_3(5)$? $r_3(5^2)$? $r_3(5^3)$? $r_3(5^{23561})$? $r_3(5^{244})$? Generalize.
6. Now analyze higher powers until you see a pattern.
7. What is $r_3(365^{1354234})$? $r_3(132444^{2352})$? $r_3(8764^{1323})$?

Answers, Simpler Problem

1. 1; the mod 3 remainder of any power of 1 is 1.
2. 2; 1; 2; 1; 2; 2; 1. Since $2 = -1 \pmod 3$, even powers of 2 have a remainder of 1, and odd powers of 2 have a remainder of 2.
3. 0; 0; the mod 3 remainder of any power of 3 is 0.
4. 1; 1; 1; 1. Since $r_3(4) = 1$, $r_3(4^k) = [r_3(4)]^k = 1^k = 1$ for any positive integer k. (Compare this answer to your answer to part 1.)
5. 2; 1; 2; 2; 1. Even powers of 5 have a remainder of 1, and odd powers of 5 have a remainder of 2. (Compare this answer to your answer to part 2. The answers are the same since $r_3(5) = 2$; $r_3(5^k) = 2^k \pmod 3$.)
6. The powers of 6 behave like the powers of 3; The powers of 7 behave like the powers of 1; The powers of 8 behave like the powers of 2; *etc*. Note that any integer is of the form $3j, 3j+1$, or $3j+2$ (that is, has a mod 3 remainder of 0, 1, or 2). So, we could express the generalization as follows.
   For any positive integer k:
   $$r_3([3j]^k) = 0,$$
   $$r_3([3j+1]^k) = 1,$$
   $$r_3([3j+2]^k) = 1 \text{ if } k \text{ is even,}$$
   and $$r_3([3j+2]^k) = 2 \text{ if } k \text{ is odd.}$$
7. $r_3(365) = 2$ and $2^{1354234} = 1 \pmod 3$; $r_3(132444) = 0$ and $0^{2352} = 0 \pmod 3$; $r_3(8764) = 1$, and $1^{1323} = 1 \pmod 3$.

"We were able to discover a method for finding the mod 3 remainder of numbers of the form a^k by systematically investigating mod 3 remainders of powers and looking for a pattern. I suggest we use the same strategy to solve the problem $r_7(216310^{65374})$. Fill out the right-hand portions of each of the following tables until you see the pattern for that table. Then move on to the next table and do the same thing. Keep doing this until you see 'the whole picture'. (Do not expect it to come quite as quickly as it did for mod 3 remainders.) At that point you are ready to solve the problem: $r_7(216310^{65374})$."

x	$r_7(x)$
2	
2^2	
2^3	
2^4	
2^5	
2^6	
2^7	
(*etc.*)	

x	$r_7(x)$
3	
3^2	
3^3	
3^4	
3^5	
3^6	
3^7	
(*etc.*)	

x	$r_7(x)$
4	
4^2	
4^3	
4^4	
4^5	
4^6	
4^7	
(*etc.*)	

x	$r_7(x)$
5	
5^2	
5^3	
5^4	
5^5	
5^6	
5^7	
(*etc.*)	

x	$r_7(x)$
6	
6^2	
6^3	
6^4	
6^5	
6^6	
6^7	
(*etc.*)	

x	$r_7(x)$
7	
7^2	
7^3	
7^4	
7^5	
7^6	
7^7	
(*etc.*)	

x	$r_7(x)$
8	
8^2	
8^3	
8^4	
8^5	
8^6	
8^7	
(*etc.*)	

x	$r_7(x)$
9	
9^2	
9^3	
9^4	
9^5	
9^6	
9^7	
(*etc.*)	

x	$r_7(x)$
10	
10^2	
10^3	
10^4	
10^5	
10^6	
10^7	
(*etc.*)	

x	$r_7(x)$
11	
11^2	
11^3	
11^4	
11^5	
11^6	
11^7	
(*etc.*)	

x	$r_7(x)$
12	
12^2	
12^3	
12^4	
12^5	
12^6	
12^7	
(*etc.*)	

x	$r_7(x)$
13	
13^2	
13^3	
13^4	
13^5	
13^6	
13^7	
(*etc.*)	

"I did it, Ant! I see what is going on. Though it is very strange, indeed. You have to look at the mod 7 remainder of the base, but the *mod 6* remainder of the exponent. I got $r_7(216310^{65374}) = r_7(3^4) = 4$. Give me a couple of others like that."

Supplementary Problem 5.4 (or has this now become an exercise?)

Find $r_7(35761^{2542})$ and $r_7(70146^{66245})$.

"The set of *nonzero* elements of $\mathbf{Z}_p$ under $\cdot_p$, where p is prime, is an example of a **group***. An important result in group theory states that if a finite group has t elements, then for each element x in the group, $x^t = 1$. Thus, $x^2 = 1 \pmod 3$ for any *nonzero* x in $\mathbf{Z}_3$ (there are two such elements) and $x^6 = 1 \pmod 7$ for any *nonzero* x in $\mathbf{Z}_7$ (there are six such elements.) However, if n is composite, $\mathbf{Z}_n$ does not form a group under $\cdot_n$. (To qualify as a group each element must have a multiplicative inverse.) But if you toss 0 and all the zero divisors out of $\mathbf{Z}_n$, the elements left form a group under mod n multiplication and then the general group theory result applies. For example, $\mathbf{Z}_9 = \{0, 1, 2, 3, 4, 5, 6, 7, 8\}$. Eliminate 0, 3, and 6 and you are left with $U_9 = \{1, 2, 4, 5, 7, 8\}$ (the units of $\mathbf{Z}_9$). U_9 has 6 elements. Hence, for each x in U_9, $x^6 = 1 \pmod 9$.

"Now had we known this general result about groups *prior* to investigating Supplementary Problem 5.3, and had we recognized $U_7 = \{1, 2, 3, 4, 5, 6\}$ as a group with respect to $\cdot_7$, we could have reasoned (without developing any tables) as follows.

---

*A multiplicative group, $(S, \cdot)$, is an algebraic structure that has the properties necessary to uniquely solve any equation of the form $a \cdot x = b$. Thus, multiplication is associative, S contains a multiplicative identity, and each element of S has a multiplicative inverse in S.

"For each element x in U_7, $x^6 = 1 \pmod 7$, so by the Super-Important Theorem:

$$\begin{aligned} r_7(216310^{65374}) &= [r_7(216310)]^{65374} && \pmod 7 \\ &= 3^{65374} && \pmod 7 \\ &= 3^{[6(10895)+4]} && \pmod 7 \\ &= (3^6)^{10895} \cdot 3^4 && \pmod 7 \\ &= 1^{10895} \cdot (3^4) && \pmod 7 \\ &= 3^4 && \pmod 7 \\ &= 4 && \pmod 7.^* \end{aligned}$$

"That is, by knowing a little general group theory, I can sometimes more easily solve problems about specific groups."

"Can't all this wait for another book, Ant?"

"Very well. Let's move on to the Football Problem."

Supplementary Problem 5.5

What is $r_5(31453^{31259})$? $r_{11}(6734521^{215424})$? $r_{25}(254304^{403})$?

Supplementary Problem 5.6

For each positive integer $n > 1$, define $\phi(n)$ to be the number of elements of U_n, the units of $\mathbf{Z}_n$. That is, $\phi(n)$ is the number of positive integers less than n which are relatively prime to n. This function is called the **Euler ϕ-function**.

1. Suppose that p is prime. What is $\phi(p)$? $\phi(p^2)$? $\phi(p^3)$? $\phi(p^k)$?
2. Suppose that p and q are distinct primes. What is $\phi(pq)$? $\phi(p^2q)$? $\phi(p^3q^2)$? $\phi(p^rq^s)$?
3. Suppose that p_1, p_2, and p_3 are distinct primes. What is $\phi(p_1p_2p_3)$? $\phi(p_1^2p_2^3p_3)$? Generalize.
4. Find a positive integer k such that $7^k = 1 \pmod{45}$. Find a positive integer k such that $31^k = 1 \pmod{1800}$. (Note: 7 is an element of U_{45} and 31 is an element of U_{1800}.)

---

*Notice that once again, it is only a remainder, $r_6(65374)$, that is needed. I expended energy unnecessarily in also computing the quotient, 10895.

Section Six

The Football Problem: Linear Diophantine Equations

I've always preferred mental exercise to the physical sort. Ponder a problem and all mankind is edified. For contemplation is one of the loftiest of human endeavors. But darting around a field while scuffling over a ball and inflicting bodily injury ranks among the most banal.

In any case, that is the argument I tried on Gnam to avoid going to a football game. It didn't work. We went to the game.

While Gnam and the other fans cheered wildly, I sat in my seat, deeply involved in thought. I have to admit it; I found the game fascinating.

Afterwards I said to Gnam: "How can it be that for so many years I didn't realize what an important role numbers play in football?"

"Well we can't all be bright," replied Gnam.

Wishing to ignore that remark, I continued: "As I recall, the last time I went to a football game, the final score was 2 to 1."

"That's impossible," said Gnam. "Given a scoring opportunity, a team can make either 2, 3, 6, 7, or 8 points. So a final score can't possibly be 2 to 1."

"Interesting." I replied. "And just which other scores are also impossible?"

Most of the way home Gnam and I talked about football scores. We decided to consider games in which the only ways to make points are from field goals (3 points each) and touchdowns with extra point (7 points each). It simplified considerably several of the problems we were posing. *Henceforth, this is the situation under consideration.*

As we walked we posed the following problems.

Which final scores for a team are possible? Clearly there is no largest one.* But is there a largest *impossible* score? And if there are only a finite number of impossible scores, just how many are there?

The final score of the game we attended was 24 to 17. Given that information only, can one deduce how many field goals and touchdowns each team made? Of course, the assumption is that all points made were from field goals and touchdowns with extra point.

In general, given a possible score, can one deduce how many touchdowns and field goals were made? Can there be more than one way to make a possible score? For which scores and how many ways?

The two of us were thinking of so many questions that Gnam started writing them down. That's not an easy thing to do when you're walking.

And then the most incredible thing happened! A man who Gnam later swore was a football star approached us and said "And just what do you think is so special about touchdowns being worth 7 points and field goals 3? Suppose touchdowns were worth 9 points and field goals 4? How would that affect the answers to your questions?"

In awe of the football hero, Gnam became tongue-tied. Meanwhile I began to fantasize that I was surrounded by football players, each one yelling out a question more intriguing than the question before it. But there was only this one player pointing at us and saying "What you *should* be considering is the situation in which field goals are worth a points each and touchdowns with extra point are worth b points each."

When we arrived home, Gnam noticed this manuscript sitting on my desk, and said: "That's a curious title for Section 6. What do any of the football problems that we've been discussing have to do with equations?"

"Gnam, habit number one to develop is this: when given a mathematical situation, translate it into the language of mathematics. For example, how would you express the question of

---

*Well, maybe there is due to time constraints, but I need a football fan to tell me what it would be. Gnam and I are assuming here that games could go on indefinitely, which is how I tend to feel they do.

whether a particular score k is a possible score in football, given that all points obtained are from field goals or touchdowns with extra point?"

"Reluctantly," replied Gnam.

So I answered my own question. "The problem is equivalent to: Given a whole number k, do there exist whole numbers x and y such that

$$7x + 3y = k?$$

"The quest for whole number (or integral) solutions to equations is a common mathematical problem and this area of study is called *Diophantine Analysis* after Diophantos of Alexandria, who is believed to have lived about 250 A.D. I have prepared a short lecture on Diophantos for inclusion in this book, Gnam. Would you like to hear it?"

"If it is interesting."

Thinking it was, I read Gnam the following.

Diophantos sought whole number solutions (and sometimes positive rational solutions) to equations. Nowadays, a **Diophantine equation** means an equation with integral coefficients* for which integral solutions are sought. What is known about Diophantos' personal life is summed up in the following puzzle, found in the Greek collection of problems in poetic form "Palatine Anthology" which was compiled about a century after Diophantos lived.

> Here you see the tomb containing the remains of Diophantos, it is remarkable; artfully it tells the measures of his life. The sixth part of his life God granted him for his youth. After a twelfth more his cheeks were bearded. After an additional seventh he kindled the light of marriage, and in the fifth year he accepted a son. Elas, a dear but unfortunate child, half of his father he was and this was also the span a cruel fate granted him, and he consoled his grief in the remaining four years of his life. By this device of numbers, tell us the extent of his life.

---

*An equation with rational coefficients is equivalent to one with only integral coefficients. You can express each coefficient as a fraction, and multiply both sides of the equation by the product of the denominators of the coefficients.

Some of Diophanto's methods for finding solutions to equations are recorded in his books, less than half of which have survived. Like his Greek predecessors, and unlike modern mathematicians, Diophantos was content to find only a few solutions to an equation, though there may be infinitely many others. Some of the problems Diophantos investigated were:

> Book III Problem 6. Find three numbers whose sum is a square and such that the sum of any two of them is a square.
>
> Book III Problem 7. Find a right triangle such that the area added to one of its perpendiculars makes 7.
>
> Book VI Problem 1. Find a right triangle such that the hypotenuse minus either of the other two sides is a perfect cube.*

A copy of Diophanto's *Arithmetica* landed in the hands of the French mathematician Pierre de Fermat nearly 1400 years after it was written. In the margins of this book, Fermat noted many of his own results, but without proofs or indications of his methods. One of his comments has since become immortalized. In 1637, opposite the Diophantine equation $x^2 + y^2 = z^2$, Fermat wrote:

> But to divide a cube into two cubes, a fourth power into two fourth powers, and in general any power whatsoever above the second into two powers of the same denomination is impossible, of which fact I have indeed discovered a marvelous proof. This margin is too narrow to hold it.

In other words, Fermat asserted that the Diophantine equation

$$x^n + y^n = z^n$$

has no nonzero solutions if $n > 2$.

The last of Fermat's remarks in the book needing proof, the conjecture was called "Fermat's Last Theorem".

For more than 300 years, Fermat's Last Theorem has enticed not only mathematicians but also laymen. Arguably the most

---

*One solution: Book III Problem 6, (80, 320, 41); Book III Problem 7, (6, 7/4, 25/4); Book VI Problem 1, (40, 96, 104)

popular of unsolved problems among nonmathematicians, erroneous "proofs" to the famous conjecture were regularly submitted by laymen hoping to win fame as well as monetary rewards. But despite efforts of even the world's greatest mathematicians, including Euler, Gauss, and Cauchy, the theorem was proved only for special cases of n. Nevertheless these efforts were not unfruitful, for from them emerged the basis of much of modern number theory and algebra.

In 1993, Princeton University professor Andrew Wiles announced that he had proved Fermat's Last Theorem. The mathematical community exploded with excitement. Worldwide, daily newspapers reported Wiles' claim. However, as of this writing, the verdict is still out on whether he has succeeded in bringing the search for a proof to an end.

"While you were talking, I figured out how many years Diophantos lived. When do we get back to a discussion of football?"

"One of the questions we posed is of particular interest. I'm calling it the Football Problem."

The Football Problem

> If a, b, and k are integers, find all pairs of integers (x, y) such that $ax + by = k$. That is, solve the Diophantine equation $ax + by = k$.

"Except for its name, what does that have to do with football?"

"The question of how a team can obtain a particular score k can be translated to the problem of finding all pairs of whole numbers (x, y) such that $7x + 3y = k$. I just generalized the situation and gave football the credit! Actually, many empirical problems give rise to linear Diophantine equations."

Exercise 6.1

The following situations suggest which linear Diophantine equations?

1. The home team scored 20 points from touchdowns with extra point and field goals.
2. I spent $325 buying shoes, at a cost of $20 per pair, and hats, at a cost of $25 per hat.
3. We need to seat 30 guests at some combination of round tables which seat 6, and square tables which seat 4. (All seats are to be occupied.)
4. The total number of points I received on my exams this semester was 560. Surprisingly, I received either 85 or 95 points on each of the exams.
5. I am dealt a bridge hand containing 19 points in aces and kings. Aces are worth 4 points and kings are worth 3 points. (There are only 4 aces and 4 kings in a bridge deck.)
6. The home team beat the visiting team by 5 points. All points scored by both teams were from field goals and touchdowns with extra point.
7. We took in $130 last night, charging an admission price of $3 for adults and $1.50 for children.

Answers, 6.1

1. $7x+3y=20$ where x is the number of touchdowns, and y is the number of field goals scored.
2. $20x+25y=325$ where x is the number of pairs of shoes purchased, and y is the number of hats purchased.
3. $4x+6y=30$ where x is the number of square tables needed and y is the number of round tables needed.
4. $85x+95y=560$ where x is the number of exams I received 85 on, and y is the number of exams I received 95 on.
5. $4x+3y=19$ where x is the number of aces I am dealt and y is the number of kings I am dealt.
6. $7x+3y=5$ where x is the difference between the number of touchdowns made by the home team and the number of touchdowns made by the visiting team and y is the difference between the number of field goals made by the home team and the number of field goals made by the visiting team. This is the only situation described in this exercise in which it makes sense for x or y to be negative. If $x=-1$ and $y=4$, then the home team scored one fewer touchdown but 4 more field goals than the visiting team.
7. $3x+1.5y=130$ which is equivalent to $6x+3y=260$, where x is the number of adult tickets sold and y is the number of children's tickets sold.

"We will now develop a technique for solving linear Diophantine equations. Consider the equation $7x + 3y = 20$ which corresponds to **situation 1** in the previous exercise. We seek all pairs of whole numbers (x, y) satisfying this equation. We are going to employ a very clever, but not uncommon, strategy used in mathematics. Sometimes one can gain insight about a problem described within the framework of one algebraic structure by looking at a 'corresponding' problem in a different algebraic structure. Here we are going to solve the Diophantine equation $7x + 3y = 20$ by first detouring through an appropriately chosen modular system."

"Now, if x and y are integers such that

$$7x+3y=20$$

then

$$r_3(7x+3y)=r_3(20).$$

Thus

$$r(7x)+r(3y)=2 \pmod 3$$

so

$$1r(x)+0=2 \pmod 3.$$

$$r_3(x)=2.$$

"Since x leaves a remainder of 2 when divided by 3, x is 2 more than a bunch of threes. That is, x is of the form $2 + 3t$, where t is an integer. If we substitute $2 + 3t$ for x in the equation $7x + 3y = 20$, we obtain:

$$\begin{aligned} 7(2+3t)+3y&=20\\ 14+21t+3y&=20\\ 3y&=6-21t\\ y&=2-7t. \end{aligned}$$

"Thus, the solution to the Diophantine equation $7x + 3y = 20$ is $\{(2+3t, 2-7t)$: t is an integer$\}$. If we let $t = 0$ we obtain the pair (2, 2). (The team scored 2 touchdowns and 2 field goals.)

"If we let $t = 1$, we obtain (5, –5), which is a solution of the equation but is meaningless in the context of the situation which gave rise to the equation. Similarly, if $t \geq 2$, the y component will be negative.

"If we substitute $t = -1$ (or any smaller t), then the x component will be negative. Hence the only pair of whole numbers (x, y) which satisfies $7x + 3y = 20$ is the pair (2, 2)."

"But why did you choose to examine the equation over $\mathbf{Z}_3$ rather than some other modular system, Ant?"

"My strategy here is to map an equation in two unknowns over the integers into an equation in one unknown over some modular system. If I examine $7x + 3y = 20$ over, say, $\mathbf{Z}_4$ I get $r(7x + 3y) = r(20) \pmod 4$ or $3r(x) + 3r(y) = 0 \pmod 4$, which is just another equation in two unknowns."*

"But why couldn't you have argued mod 7?"

"I could have. Let's do it."

If x and y are integers such that

$$7x + 3y = 20$$

then $r(7x + 3y) = r(20) \pmod 7$

so $r(7x) + r(3y) = 6 \pmod 7$

thus we have an equation in one unknown:

$$3r(y) = 6 \pmod 7$$

$$r(y) = 2 \pmod 7.$$

"Therefore, y is of the form $2 + 7t$, where t is an integer. Substituting $2 + 7t$ for y, we obtain $7x + 3(2 + 7t) = 20$. From this we get $7x + 6 + 21t = 20$ and $x = 2 - 3t$. Thus, the solution to the Diophantine equation $7x + 3y = 20$ is $\{(2 - 3t, 2 + 7t)$: t is an integer$\}$."

"Just what I suspected! You get a different answer! The first answer was $(2 + 3t, 2 - 7t)$. The +'s and –'s are reversed!"

"The two sets just look different, Gnam. But really they describe precisely the same set of ordered pairs, because subtracting, say, a bunch of threes is like adding the opposite of that bunch of threes. When you substitute –1 for t in $(2 - 3t, 2 + 7t)$, you get $(5, -5)$, which is the same solution that you get when you substitute 1 for t in $(2 + 3t, 2 - 7t)$. In the following table, I used the value of t in the left column to generate the values of $2 + 3t$ and $2 - 7t$, but I used its opposite value in the right column to generate the values of $2 - 3t$ and $2 + 7t$."

	x	y
t	$2+3t$	$2-7t$
–3	–7	23
–2	–4	16
–1	–1	9
0	2	2
1	5	–5

x	y	
$2-3t$	$2+7t$	t
–7	23	3
–4	16	2
–1	9	1
2	2	0
5	–5	–1

*After you've gotten your feet wet, come back to the equation $7x + 3y = 20$ and try using, say, a mod 14 argument on it. What happens, and *why*?

"Substituting –3 for t on the left yields the same (x, y) as substituting 3 for t on the right. In general, $(2 + 3[-t], 2 - 7[-t]) = (2 - 3t, 2 + 7t)$. So, although a specific value of t (such as 1) might generate different pairs in the two sets [here, (5, –5) and (–1, 9)], in totality the two sets describe the same pairs of numbers because t can be negative as well as nonnegative."

"I see something else, Ant. The values of x increase by 3 and the values of y decrease by 7. (That is, unless you read the tables from the bottom going up. Then the values of x decrease by 3 and the values of y increase by 7.) But 7 was the coefficient of x and 3 was the coefficient of y in your equation. Everything got reversed!"

"That's an important observation, Gnam, one that I'd like to come back to later. Now, the graph of the equation $7x + 3y = 20$ is a line which passes through infinitely many points whose coordinates are both integers, but only one point whose coordinates are both whole numbers, namely (2, 2)."

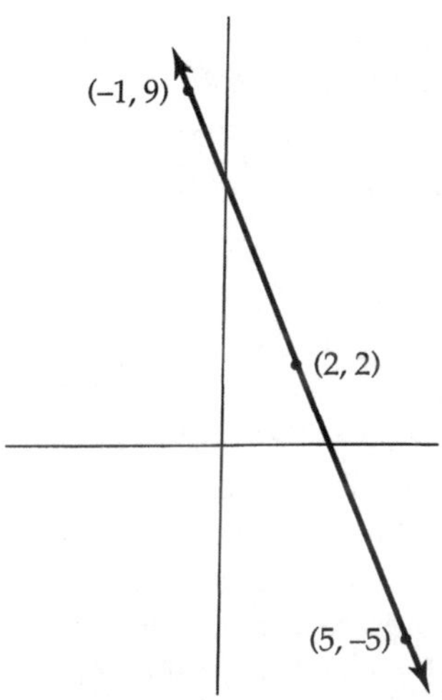

No matter which solution you start with, if you add a multiple of 3 to the x coordinate and subtract the same multiple of 7 from the y coordinate, you will obtain another solution. Indeed, we could have described the solution to the Diophantine equation $7x + 3y = 20$ as $\{(-1 + 3t, 9 - 7t)$: t is an integer$\}$. (Or, if you prefer, $\{(-1 - 3t, 9 + 7t)$: t is an integer$\}$.)

"Let's now consider the equation suggested by **situation 2** of the last exercise: $20x + 25y = 325$. Dividing this equation by 5, we obtain the equivalent equation $4x + 5y = 65$. I will use a mod 5 argument to eliminate the y.

If x and y are integers such that

$$4x + 5y = 65$$

then
$$r_5(4x) = r_5(65)$$

so
$$4r(x) = 0 \quad (\text{mod } 5)$$

hence
$$r(x) = 0 \quad (\text{mod } 5).$$

"Thus, x is a multiple of 5. That is, x is of the form $5t$ where t is an integer. Substituting $5t$ for x in the equation $4x + 5y = 65$ we obtain:

$$4(5t) + 5y = 65$$
$$5y = 65 - 20t$$
$$y = 13 - 4t.$$

"So the solution to the Diophantine equation $4x + 5y = 65$ is $\{(5t, 13 - 4t)$: t is an integer$\}$. But we are seeking only those solutions which are pairs of whole numbers. Now:

if $5t \geq 0$ then $t \geq 0$

and
if $13 - 4t \geq 0$, then $13 \geq 4t$ so $t \leq 13/4$.

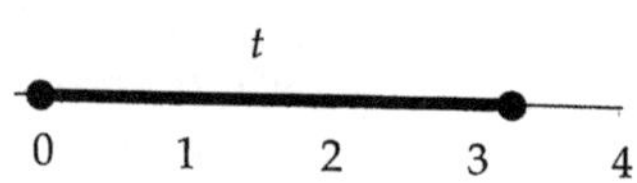

"The only integers t which satisfy both conditions are 0, 1, 2, and 3. If we let $t = 0$, we obtain the solution $(5 \cdot 0, 13 - 4 \cdot 0) = (0, 13)$. (I bought no pairs of shoes, and 13 hats.) If we let $t = 1$, we obtain the solution (5, 9). (I bought five pairs of shoes and 9 hats.) If we let $t = 2$ we obtain (10, 5) (10 pairs of shoes and 5 hats) and if we let $t = 3$ we obtain (15, 1) (15 pairs of shoes and 1 hat). Thus, the solutions which are whole number pairs are (0, 13), (5, 9), (10, 5), and (15, 1). The values of x increase by 5 while the values of y decrease by 4. Recall that the equation we solved is $4x + 5y = 65$. The graph of this equation passes through 4 points in the first quadrant whose coordinates are both whole numbers. "

Exercise 6.2

> Solve the Diophantine equation $4x + 5y = 65$ by using a mod 4 argument instead of a mod 5 argument. Does your solution contain the same pairs of whole numbers?

"Here we have equations with coefficients which are integers, solutions which are integers, and yet we detour through the modular systems to solve the equations. The method that the 19th century mathematician E. E. Kummer used to try to prove Fermat's Last Theorem was to rewrite the Diophantine equation $x^n + y^n = z^n$ as $y^n = z^n - x^n$ and then factor the right hand side into a product of linear polynomials, not with integral coefficients, but with complex coefficients! It is fascinating that one can gain information about a problem by examining it in a context different from the one that...."

"Stop! Let's continue solving these equations."

"The Diophantine equation corresponding to **situation 3** is $4x + 6y = 30$.

$$4x + 6y = 30$$

$\Leftrightarrow$
$$2x + 3y = 15.$$

Thus
$$2r(x) = 0 \pmod 3$$
$$r(x) = 0 \pmod 3.$$

"So, $x = 3t$. Now, $2(3t) + 3y = 15$ implies that $y = 5 - 2t$. Thus the solution to the Diophantine equation $2x + 3y = 15$ is $\{(3t, 5 - 2t)$: t is an integer$\}$. Once again, we seek only those solutions which are pairs of whole numbers.

"If $3t \geq 0$ then $t \geq 0$ and if $5 - 2t \geq 0$ then $t \leq 5/2$.

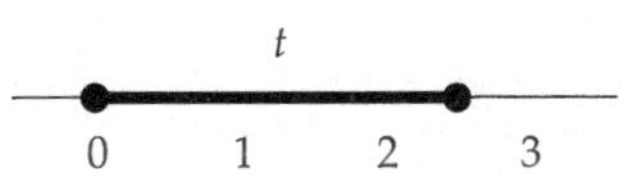

Answer, 6.2
If $4x + 5y = 65$, then $r_4(4x + 5y) = r_4(65)$. So $r_4(y) = 1$. Thus $y = 4t + 1$. Solving for x, we obtain: $4x + 5(4t + 1) = 65$, so $x = 15 - 5t$. Thus, the solution to the Diophantine equation is $\{(15 - 5t, 1 + 4t)$: t is an integer$\}$. You obtain the same whole number pairs (0, 13), (5, 9), (10, 5), and (15, 1) by letting $t = 3, 2, 1$, and 0, respectively.

"Thus, t is 0, 1, or 2. These values of t produce the pairs (0,5), (3, 3), and (6, 1). (We can use 5 round tables, 3 square and 3 round tables, or 6 square and 1 round table.)"

"Before continuing your analysis of the situations, let me try solving a few equations, Ant."

Exercise 6.3

> Find all pairs of whole numbers (x, y) satisfying the following equations.
>
> 1. $8x + 5y = 57$
> 2. $14x + 22y = 262$
> 3. $13x + 12y = 59$
> 4. $3x + 5y = 47$

"I just want to reiterate one observation you made earlier before we analyze situations 4 through 6, Gnam. Look at the solutions we have obtained for the equations corresponding to the situations 1 though 3:

The solution to $7x + 3y = 20$ is $\{(2 + \mathbf{3}t, 2 - \mathbf{7}t)$: t is an integer$\}$.
The solution to $4x + 5y = 65$ is $\{(\mathbf{5}t, 13 - \mathbf{4}t)$: t is an integer$\}$.
The solution to $2x + 3y = 15$ is $\{(\mathbf{3}t, 5 - \mathbf{2}t)$: t is an integer$\}$.

"If you want to avoid doing unnecessary algebra, notice the relationship between the coefficients of t, and the coefficients of x and y in the *reduced* equations being solved. If you know *one* solution, then you know the form of the others. I'll demonstrate this in the next example."

Answers, 6.3

1. If $8x + 5y = 57$ then $3r_5(x) = 2 \pmod 5$, so $r_5(x) = 4$. Hence $x = 4 + 5t$. Substituting $4 + 5t$ for x in the equation $8x + 5y = 57$, we obtain $y = 5 - 8t$. Now, $4 + 5t \geq 0 \Leftrightarrow t \geq -4/5$ and $5 - 8t \geq 0 \Leftrightarrow t \leq 5/8$. Hence the only whole number solution is the one produced by $t = 0$, namely, the solution (4, 5).
2. The equation $14x + 22y = 262$ has the same solutions as $7x + 11y = 131$. Using a mod 11 argument, we obtain $r_{11}(x) = 3$, so $x = 3 + 11t$. Hence $y = 10 - 7t$. $x \geq 0 \Leftrightarrow t \geq -3/11$ and $y \geq 0 \Leftrightarrow t \leq 10/7$. Thus the whole number solutions are the ones produced by $t = 0$, (3, 10), and $t = 1$, (14, 3).
3. $y = 6 + 13t$ and $x = -1 - 12t$. $y \geq 0 \Leftrightarrow t \geq -6/13$, and $x \geq 0 \Leftrightarrow t \leq -1/12$. There is no whole number t satisfying both inequalities. Hence the equation $13x + 12y = 59$ has no whole number solutions.
4. There are three whole number pairs that satisfy the equation: (14, 1), (9, 4), and (4, 7).

Situation 4 Find all pairs of whole numbers (x, y) such that $85x + 95y = 560$. Now, x and y are integers such that

$$85x + 95y = 560$$

$$\Leftrightarrow \quad 17x + 19y = 112$$

so

$$\begin{aligned} 2r(y) &= r(112) \quad (\text{mod } 17) \\ 2r(y) &= 10 \quad (\text{mod } 17) \\ r(y) &= 5 \quad (\text{mod } 17). \end{aligned}$$

"So, $y = 5 + 17t$. If we substitute 5 (rather than $5 + 17t$) for y in $17x + 19y = 112$ we obtain $x = 1$. Hence, one solution is $(1, 5)$. Thus, the solution to the Diophantine equation $17x + 19y = 112$ is $\{(1 - 19t, 5 + 17t)$: t is an integer$\}$. The only solution whose components are both whole numbers is $(1, 5)$. (I received 1 eighty–five and 5 ninety–fives on my exams.)"

"Wait! There are lots of integers y such that $r(y) = 5$ (mod 17). Suppose we had chosen, say, -12 instead of 5 for y. What happens then?"

"If you substitute -12 for y in the equation $17x + 19y = 112$, you obtain $x = 20$. So another way to describe the solution to the equation is $\{(20 - 19t, -12 + 17t)$: t is an integer$\}$. Here the only value of t which generates a pair of whole numbers is $t = 1$, which yields $(1, 5)$ again. Indeed, if (x_0, y_0) is any solution of the Diophantine equation $17x + 19y = 112$, $\{(x_0 - 19t, y_0 + 17t)$: t is an integer$\}$ describes the pairs of integers which satisfy the equation."

"In other words, I have my choice of any one of infinitely many ways to describe the solution to an equation. I like that! With that much choice, how can I miss?"

"Well, of course in this situation there is only one pair, $(1, 5)$, whose components are whole numbers. But any one of those infinitely many correct choices will contain this same whole number pair."

Situation 5 Find all pairs of whole numbers (x, y), $0 \leq x \leq 4$, and $0 \leq y \leq 4$ such that $4x + 3y = 19$.

$$4x + 3y = 19$$

so $$r(x) = 1 \pmod 3$$

so $$x = 1 + 3t.$$

"Substituting 1 for x in $4x + 3y = 19$, we obtain $y = 5$. So the solution to the Diophantine equation $4x + 3y = 19$ is $\{(1 + 3t, 5 - 4t)$: t is an integer$\}$. The whole number solutions are (1, 5) and (4, 1). The solution (1, 5) does not satisfy the conditions of the problem. So the only solution to situation 5 is 4 aces and 1 king."

Situation 6 Find all pairs of *integers* (x, y) such that $7x + 3y = 5$.

$$7x + 3y = 5$$

so $$r(x) = 2 \pmod 3.$$

If $x = 2$ then $y = -3$. The solution is $\{(2 + 3t, -3 - 7t)$: t is an integer$\}$.

"Note that in this situation, it makes sense to have negative components. For example, when $t = 2$ we obtain the solution $(8, -17)$. This means that the home team made 8 more touchdowns and 17 fewer field goals than the visiting team, and thus won by 5 points."

Situation 7 Find all pairs of whole numbers (x, y) such that $6x + 3y = 260$.

$$6x + 3y = 260$$

so $$r_3(6x + 3y) = r_3(260)$$

$$0 = 2 \pmod 3.$$

"What an interesting situation, Ant! It isn't often that 0 gets to be equal to 2!"

"It doesn't here either! What we have shown is that the Diophantine equation $6x + 3y = 260$ has no solutions. The left hand side of the equation is a multiple of 3 but the right hand side isn't! (Maybe you shouldn't trust us to handle the admissions.)

"Here are some other examples of linear Diophantine equations which have no solutions."

$5x + 10y = 121$	5 is not a factor of 121
$8x + 6y = 253$	2 is not a factor of 253
$35x + 14y = 701$	7 is not a factor of 701

"In general, if $j \mid a$ and $j \mid b$ then $j \mid (ax + by)$. So if there is some integer j which divides both a and b, but doesn't divide k, the Diophantine equation $ax + by = k$ has no solutions. This means that its graph passes through no points whose coordinates are both integers.

"On the other hand, if $\gcd(a, b)$ does divide k, then our first step in solving the Diophantine equation $ax + by = k$ is to divide a, b, and k by $\gcd(a, b)$, obtaining an equivalent equation in which the coefficients of x and y are relatively prime. Whenever the coefficients of x and y are relatively prime, then, independent of the value of the constant term, the Diophantine equation does have solutions*. In this case, the graph of the equation passes through infinitely many points in the plane whose coordinates are both integers (but possibly only a finite number whose coordinates are both whole numbers, indeed possibly none)."

Theorem

> The Diophantine equation $ax + by = k$ has solutions if and only if $\gcd(a, b) \mid k$. In particular, if $\gcd(a, b) = 1$ there are integers x and y such that $ax + by = 1$.

"We'll add a small twist to our final example."

Example

> Find all pairs of whole numbers (x, y) such that
> $$22x - 622y = 3562.$$

---

*Here's a proof when a and b are both positive. Suppose that $\gcd(a, b) = 1$. Either $a = b$ or $a \neq b$. If $a = b$ then $a = 1$ and the Diophantine equation $x + y = k$ has solutions. If $a \neq b$, say, $a < b$, then $/a \pmod b$ exists, and so there is an integer x such that $ax = r(k) \pmod b$. So $b \mid [r_b(k) - ax]$. Thus there is an integer y such that $ax + by = r_b(k)$. But $r_b(k) = k - qb$ for some integer q. Hence $ax + b(q + y) = k$. The ordered pair of integers $(x, q + y)$ is a solution of the equation.

Gcd(22, 622) = 2 and $2 \mid 3562$. Dividing both sides of the equation by 2, we obtain the equivalent equation $11x - 311y = 1781$. Now, if x and y are integers such that $11x + (-311y) = 1781$, then

$$\begin{aligned} r_{11}(-311y) &= r_{11}(1781) \\ -3r(y) &= 10 = -1 \quad (\text{mod } 11) \\ 3r(y) &= 1 \quad (\text{mod } 11) \\ r(y) &= 4 \quad (\text{mod } 11). \end{aligned}$$

If $y = 4$ then $x = 275$. Thus the solution to the Diophantine equation $11x - 311y = 1781$ is $\{(275 - (-311t), 4 + 11t): t \text{ is an integer}\} = \{(275 + 311t, 4 + 11t): t \text{ is an integer}\}$. Now, $275 + 311t \geq 0$ when $t \geq -275/311$ and $4 + 11t \geq 0$ when $t \geq -4/11$. So *every* whole number t generates a solution whose components are whole numbers. Thus, the set of pairs of whole numbers satisfying the equation $11x - 311y = 1781$ is $\{(275 + 311t, 4 + 11t): t \text{ is a whole number}\}$. As you can see from the diagram here, the graph of this line passes through *infinitely* many points which have whole number coordinates.

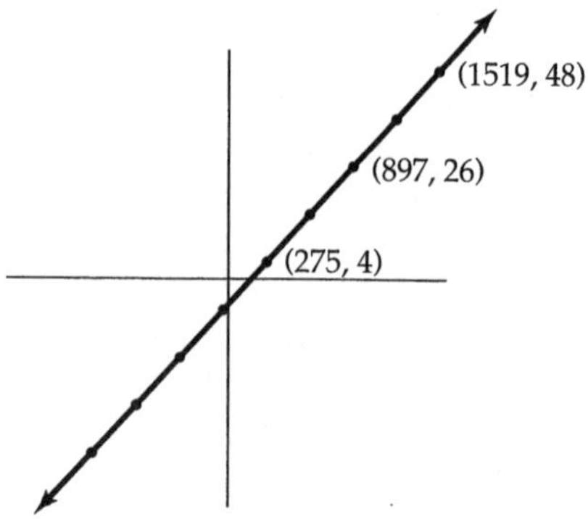

Rough graph of $11x - 311y = 1781$

"In summary, to solve the Diophantine equation $ax + by = k$, divide a, b, and k by $g = \gcd(a, b)$, so that the resulting coefficients of x and y are relatively prime. (If g is not a factor of k, then the equation has no solutions.) Then find *any one* solution (x_0, y_0) of the equation. The solution to $ax + by = k$ is $\{(x_0 + (b/g)t, y_0 - (a/g)t): t \text{ is an integer}\}$. Sometimes you may be able to 'eyeball' a solution. For example, if it is apparent that (1, 3) is a solution of $2x + 5y = 17$, then you can sidestep use of the modular systems, and just note that the solution to the Diophantine equation is $\{(1 + 5t, 3 - 2t): t \text{ is an integer}\}$."

Exercise 6.4

Solve the following Diophantine equations.

1. $11x + 310y = 1781$
2. $21x + 28y = 25$
3. $9x + 13y = 47$
4. $31x + 56y = -1$
5. $8x + 6y = 48$
6. $101x + 120y = 3$
7. $97x - 179y = 1$
8. $53x - 17y = 177$

Exercise 6.5

Find all pairs of whole numbers (x, y) such that:

1. $11x + 311y = 1781$
2. $16x + 93y = 891$
3. $21x + 36y = 372$
4. $29x + 33y = 5814$
5. $17x - 23y = 473$

Answers, 6.4

If (x_0, y_0) is any solution of the Diophantine equation $ax + by = k$, where a and b are relatively prime, then the solution to the equation is either $\{(x_0 + bt, y_0 - at)$: t is an integer$\}$ or, if you prefer, $\{(x_0 - bt, y_0 + at)$: t is an integer$\}$. (You are asked to prove this in Problem 6.1.) Thus, your solution may appear to be different than mine when, in fact, both are correct for our solutions are equivalent.

1. $\{(21 + 310t, 5 - 11t)$: t is an integer$\}$
2. No solutions since gcd(21, 28) is not a factor of 25.
3. $\{(-2 - 13t, 5 + 9t)$: t is an integer$\}$
4. $\{(-47 - 56t, 26 + 31t)$: t is an integer$\}$
5. $\{(3t, 8 - 4t)$: t is an integer$\}$
6. $\{(-57 - 120t, 48 + 101t)$: t is an integer$\}$
7. $\{(24 + 179t, 13 + 97t)$: t is an integer$\}$
8. $\{(12 + 17t, 27 + 53t)$: t is an integer$\}$

Answers, 6.5

1. The solutions are of the form $(-36 + 311t, 7 - 11t)$. There is no t which produces whole number components.
2. The solutions are of the form $(15 + 93t, 7 - 16t)$. The only t which produces whole number components is 0, obtaining (15, 7).
3. The solutions are of the form $(16 + 12t, 1 - 7t)$. Letting $t = 0$ we obtain (16, 1); letting $t = -1$ we obtain (4, 8).
4. The solutions are of the form $(180 + 33t, 18 - 29t)$. The integers t such that $-5 \leq t \leq 0$ produce the solutions (180, 18), (147, 47), (114, 76), (81, 105), (48, 134) and (15, 163).
5. $\{(40 + 23t, 9 + 17t)$: t is a whole number$\}$.

Exercise 6.6

Find all whole number pairs (x, y) such that $\frac{2x}{9}+\frac{3y}{5}=17$.

Exercise 6.7

Blair and Brent set up a lemonade stand charging 35 cents for large cups and 20 cents for small cups. At the end of the day they made $4.85 (not counting expenses). How many large cups and how many small cups of lemonade did they sell?

Exercise 6.8

For each of the following, find a linear Diophantine equation which has the indicated solution.

1. $\{(5-2t, 4+7t)$: t is an integer$\}$
2. $\{(5-2t, 4-7t)$: t is an integer$\}$
3. $\{(5-2t, 4+6t)$: t is an integer$\}$
4. Two of its solutions are (5, 1) and (7, –2); two of its solutions are (2, –1) and (5, 3); two of its solutions are (7, 5) and (13, 2).

Answers, 6.6
(9, 25), (36, 15), and (63, 5)

Answer, 6.7
There are three possibilities: (11, 5), (7, 12), and (3, 19).

Answers, 6.8

1. $7x+2y=43$.
2. $7x-2y=27$.
3. There is no linear Diophantine equation with precisely this set of solutions. The set describes some of the solutions of the equation $6x+2y=38$ (which is equivalent to $3x+y=19$), but not all of them. For example, (6, 1) is a solution of $6x+2y=38$, but is not included in the set.

   In general, $\{(x_0-bt, y_0+at)$: t is a whole number$\}$, where $a \neq 0$ and $b \neq 0$, is the solution of a linear Diophantine equation $\Leftrightarrow$ a and b are relatively prime. (Investigate what happens if either or both coefficients of t are 0.)
4. $3x+2y=17$ since if (5, 1) and (7, –2) are two solutions then the x components are increasing by multiples of 2 while the y components are decreasing by the same multiples of 3. So the equation is of the form $3x+2y=k$. To find k, substitute either given point into $3x+2y$, obtaining $k=17$; $4x-3y=11$; $3x+6y=51$ or $x+2y=17$.

Exercise 6.9

Make up a linear Diophantine equation which has solutions, but none of the solutions are pairs of whole numbers (that is, every solution has at least 1 negative component).

Exercise 6.10

You are the conductor for a one hundred-member high school band. One morning when the students arrive for practice, you are feeling too lazy to count the number of students who showed up, so you tell them to line up in rows of 20. They do so and although you forget to notice how many rows of 20 there are, you do see 3 students left in a row by themselves. Then you tell them to line up in rows of 13. They do so and once again you forget to count how many rows of 13 they make, but you do note that there are 5 students in a row by themselves. How many students showed up for practice?

Supplementary Problem 6.1

1. Show that if (x_0, y_0) is any solution of the Diophantine equation $ax + by = k$ then for each integer t, the ordered pair $(x_0 + bt, y_0 - at)$ is a solution of the equation.
2. Suppose that (x_0, y_0) and (x_1, y_1) are both solutions of the Diophantine equation $ax + by = k$, where $\gcd(a, b) = 1$. Prove that there exists an integer t such that $x_1 = x_0 + bt$ and $y_1 = y_0 - at$.

Thus, if (x_0, y_0) is a solution of the Diophantine equation $ax + by = k$, where $\gcd(a, b) = 1$, then (x_1, y_1) is a solution of this equation if and only if there exists an integer t such that $x_1 = x_0 + bt$ and $y_1 = y_0 - at$.

Answers, 6.9
There are infinitely many possibilities. One is: $5x + 2y = 1$.

Answer, 6.10
83. If x is the number of students in attendence, then $x \leq 100$, and there exist whole numbers k and j such that $x = 20k + 3$ and $x = 13j + 5$. So, $20k + 3 = 13j + 5$. If you solve this Diophantine equation you obtain $r_{13}(k) = 4$. If $k = 4$ then $x = 83$. No other k of the form $4 + 13t$ yields a whole number x less than or equal to 100.

Supplementary Problem 6.2

1. Assuming that all points are made from touchdowns with one extra point and field goals, how many impossible scores are there, and what is the largest impossible score?
2. Suppose that a and b are relatively prime positive integers. How many positive integers k are there such that the Diophantine equation $ax + by = k$ has no solutions whose components are both whole numbers? What is the largest such k?
   One possible strategy: Make a table with four columns labelled a, b, the number of impossible scores, and the largest impossible score. Choose small, relatively prime values for a and for b and figure out what the largest impossible score and the number of impossible scores are for that a and b. To maximize the likelihood you'll spot a pattern, you should be systematic in your choices for a and b. Rather than allow *both* of them to vary from case to case, "sit" on one, say some fairly small a, and systematically vary the other, b, until you know the formulas for that choice of a. Then increase a by 1 and repeat the procedure.
3. Suppose that there are 35 impossible scores in a game with two scoring opportunities, one worth a points and the other worth b points, where a and b are both larger than 2 and are relatively prime. Determine the largest impossible score, and the values of a and b.

Section Seven

Forward-Looking and Right-Thinking Male Chauvinist Chess

I was deeply engrossed in the Football Problem when Gnam stormed into our house.

"What an interesting chess club meeting we had this afternoon!" exclaimed Gnam. "A man whom none of us had ever seen before interrupted the chairperson and gave a speech in which he claimed that chess was subverting the basic values of our society!"

"How did he arrive at that conclusion?" I asked.

"He said that games are far more insidious than argument, or even propaganda. He said that if you enjoy a game you tend to accept its basic premises without realizing their implications."

At that point Gnam startled me by climbing on top of my desk and, with arms spread wide, orated in an imitation, I presume, of the intruder.

"Look at the basic premises of chess," Gnam said. "Do you realize that the most powerful chess piece on the board is a female! The Queen! Now you may respond that the King is the most valuable piece, and the Queen's only purpose is to protect him, but I ask you: Is it proper and fitting for a woman to protect a man in battle? Think of what this does to the self-images of the young men who play chess. Why, I wouldn't be surprised if chess were the principal cause of the feminist movement. My friends, in order to preserve traditional family values, the Queen must be eliminated from the game of chess!

"Furthermore, don't you find it suspicious that the Russians have always been masters of chess? It is not far-fetched, my friends, to suppose that chess was a leftist plot. No, not at all. Why every piece on the board is allowed to move to the left. In a forward-looking right-thinking society like our own, we should only allow the pieces to move forward and to the right."

"What happened then?" I asked, helping Gnam down from my desk. "Did anyone advise this man that his views were anachronistic?"

"Our chairperson, who tries to consider an idea on its merits rather than its source, suggested that we adopt as our game for the week a version of chess in which there is no Queen and the other pieces are allowed to move only forward and to the right. She called it 'Forward-Looking and Right-Thinking Male Chauvinist Chess'.

"Of course removing the Queen also necessitated removing the Queen's pawn, and narrowing the board to 7 squares rather than 8. To compensate for the lost space, we decided to increase the length of the board from 8 squares to 9."

"What other modifications were made?"

"That a chess piece be able to move off the board and reappear on the square it would have reached had a new board been placed next to the original one. Here, I'll show you what I mean."

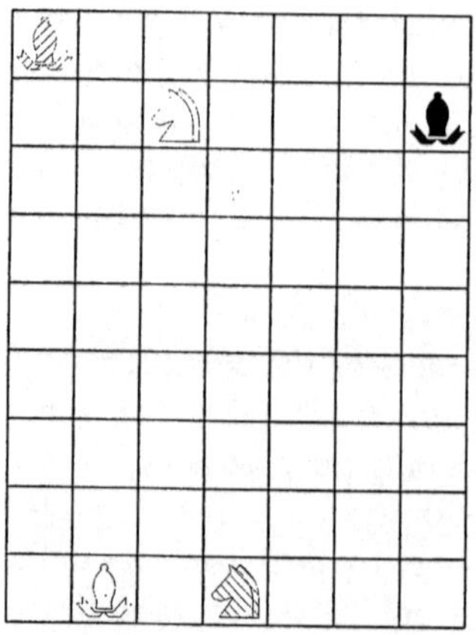

If the black bishop (♝) moves one square right-diagonally, he lands at the crossed bishop (♗). If he moves two squares right-diagonally, he lands at the gray bishop (♗). If the white knight (♘) moves one square to the right and two squares up, he lands at the crossed knight (♞).

"Aha! Chess on a doughnut! For, in effect the right and left edges are attached as are the bottom and top edges. This sounds like an interesting game." I said. "What happened?"

"We decided to give it a try. But it soon became evident that we needed the services of a mathematician to figure out which squares could be occupied by the different chess pieces and, ah, I mentioned that you were an amateur number theorist, and, well, you said you owed me one for so patiently teaching you how to program the VCR."

Of course I agreed to help Gnam investigate this modified version of chess. And that is how I came to write this section of the book. I suggested to Gnam that we not limit ourselves to an investigation of a 7 by 9 board, but rather consider moves on an m by n board. Gnam reluctantly agreed. I also suggested that we name each square of the board with ordered pairs of numbers to help us keep track of where the pieces were. Gnam said that was okay if that's what I wanted to do. I then suggested that we invite you to participate with us in our investigation. Gnam *liked* that suggestion.

"Let's name each square on an m by n board as shown in the following picture."

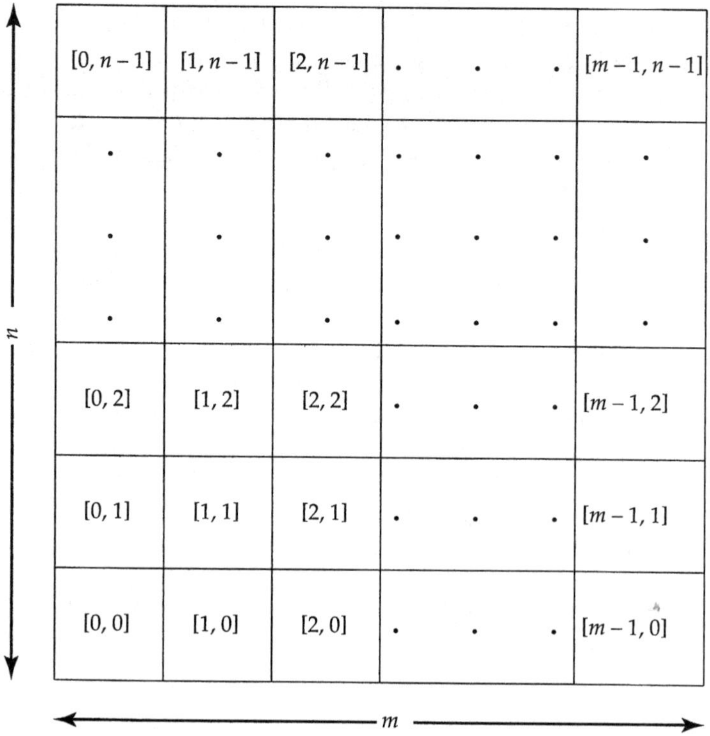

"Now that the board is nicely labelled, tell me which squares my knights can occupy, Ant."

"Before analyzing the moves of men on a two dimensional board, let's examine them in a *simpler* situation, namely on one dimensional boards; that is, on m by n boards where $n = 1$. (This type of exemplary reasoning is indicative of a highly trained mind and is worthy of emulation, Gnam.)"

Exercise 7.1

Suppose that a man is on the square [5, 0] on a 13 by 1 board. (See picture below). On what square will he land if he moves to the right:

1. 4 squares? **2.** 8 squares? **3.** 12 squares?
4. 27 squares? **5.** 130 squares? **6.** 1326399 squares?

[0, 0]	[1, 0]	[2, 0]	[3, 0]	[4, 0]	[5, 0]	[6, 0]	[7, 0]	[8, 0]	[9, 0]	[10, 0]	[11, 0]	[12, 0]

13×1 board

"The first component of the solution to part 1 is the solution of the equation $5 + 4 = k \pmod{13}$. But the equation corresponding to part 4 is not $5 + 27 = k \pmod{13}$, since 27 is not an element of $\mathbf{Z}_{13}$. However, every time a man moves 13 squares on this board he returns to his starting position. So it is the remainder when 27 is divided by 13 that is relevant. Hence, the equation corresponding to part 4 is $5 + r_{13}(27) = k \pmod{13}$.

"When analyzing one dimensional boards, let's employ the following abbreviations."

Notation (one dimensional boards)

The square $[k, 0]$ will be referred to as the square $[k]$; the m by 1 board will be referred to as just the m board, and we will use $< r >$ to denote the move 'r squares to the right'.

"There is no point to having a move larger than 12 on a 13 board, Ant. For example, moving 15 squares has the same effect as moving only 2, and it's much less trouble!"

"Then, let's restrict ourselves to moves $< r >$ where r is less than the number of squares on the board."

Answers, 7.1

1. [9, 0] 2. [0, 0] 3. [4, 0]
4. [6, 0] 5. [5, 0] 6. [1, 0]

Exercise 7.2

Given a 13 board and a man who moves $< 3 >$, on what square must the man start if, after moving once, you want him to land on:

1. 7? 2. 2? 3. 0?

"Anybody out there experiencing a feeling of déjà vu? A 13 board is nothing but a 13-hour clock stretched horizontally! Earlier you asked us what time it is on a 13-hour clock if in 3 hours it will be 7, 2 or 0, respectively."

"Then you might want to picture clocks as you answer the following questions."

Exercise 7.3

1. A man starts at [4] on a 9 board, makes one move to the right, and lands on the square [2]. What is the size of his move?
2. A man starts at [5] on a 9 board, moves $< r >$ twice, and lands on the square [8]. What is the size of his move?
3. A man starts at [10] on a 13 board, moves $< r >$ five times, and lands on the square [1]. What is the size of his move?
4. A man starts at [5], makes one move of $< 14 >$, and lands on [2]. What is the size of the board?
5. A man starts at [1] makes one move of $< 14 >$, and lands on [0]. What is the size of the board?

Answers, 7.2

Let x be the number of the square the man starts on.

1. If $x + 3 = 7 \pmod{13}$ then $x = 4$.
2. If $x + 3 = 2 \pmod{13}$ then $x = 12$.
3. If $x + 3 = 0 \pmod{13}$ then $x = 10$.

Answers, 7.3

1. $4 + r = 2 \pmod 9$ so $< r > = <7>$.
2. $5 + 2r = 8 \pmod 9$ so $< r > = <6>$.
3. $10 + 5r = 1 \pmod{13}$ so $< r > = <6>$.
4. $5 + 14 = 2 \pmod n$ so $n = 17$.
5. $1 + 14 = 0 \pmod n$, so $n = 15$. If you allow moves larger than the width of the board, then n could also be 5 or 3 (the factors of 15 larger than 1) since each is also a solution to $1 + r(14) = 0 \pmod n$.

"Suppose that a man with move < 2 > starts at [0] on an 8 board. Then he successively lands on [2], [4], [6], [0], [2], [4], [6], [0], … We call the set {0, 2, 4, 6} the orbit of the move < 2 > on the 8 board. The **orbit of a move** is just the set of squares that the man is able to occupy through successive moves, starting at [0]. Clearly, the squares that can be reached may depend on the starting position. *However, unless otherwise stated, we will start all men at [0].* The **size of an orbit** is the number of squares in the orbit. So, for example, the size of the orbit of < 2 > on an 8 board is 4."

Exercise 7.4

1. What are the orbits of each of the following moves, and what are their sizes?
   a. < 3 > on a 15 board
   b. < 6 > on an 8 board
   c. < 6 > on a 9 board
   d. < 10 > on a 12 board
   e. < 5 > on a 7 board
   f. < 6 > on a 25 board
   g. < 15 > on a 24 board
2. Continue examining specific cases until you can answer: What is the size of the orbit of $< r >$ on an m board?

Theorem 7.1

The size of the orbit of $< r >$ on an m board is $m / \gcd(m, r)$.

"A natural question to pose next is 'which squares are in the orbit of the move $< r >$ on an m board?'"

"We already figured that out on the clock, Ant. They're the mod m multiples of r. You just find the largest common factor of

---

Answers, 7.4

1. a. {0, 3, 6, 9, 12}; size 5
   b. You land on the squares in the order 6, 4, 2, and 0. But we will conventionally list the elements in the orbit in numerical order: {0, 2, 4, 6}; size 4
   c. {0, 3, 6}; size 3
   d. {0, 2, 4, 6, 8, 10}; size 6
   e. {0, 1 ,2 , 3, 4, 5, 6} (the entire board); size 7
   f. The entire board; size 25
   g. {0, 3, 6, 9, 12, 15, 18, 21}; size 8
2. See Theorem 7.1.

the move and the board size and then if a square is a multiple of that factor, it's in. Otherwise it's out!"

"There is a slightly more elegant way of expressing that idea."

Theorem 7.2

> The square $[k]$ is in the orbit of the move $< r >$ on an m board if and only if $\gcd(r, m) \mid k$. In particular, if $\gcd(r, m) = 1$, then the orbit of $< r >$ is the entire board.

Exercise 7.5

1. Is [5] in the orbit of $< 15 >$ on an 18 board?
2. Is [6] in the orbit of $< 8 >$ on a 10 board?
3. Is [3] in the orbit of $< 5 >$ on a 13 board?

"We will now investigate the question of how many times you must move $< r >$ on an m board to reach the square $[k]$. Consider the move $< 4 >$ on an 11 board. Every eleven times you move $< 4 >$ you land back on [0] (or whichever square you started on) because moving $< 4 >$ eleven times has the same effect as moving $< 11 >$ four times, and that gets you nowhere! Alternately, the size of the orbit of $< 4 >$ on an 11 board is $11/\gcd(11,4) = 11$. So every 11 times you move $< 4 >$ you return to [0]."

"So what's the point, Ant?"

"The point is that if the square $[k]$ is in the orbit of $< 4 >$ on an 11 board, it is possible to reach $[k]$ in fewer than 11 moves. Now, we know that the square [9] is in the orbit of $< 4 >$ on an 11 board, since $\gcd(4, 11) \mid 9$. So there is some x in $\mathbf{Z}_{11}$ such that if you make x moves of $< 4 >$ you will land on [9]. To find x we solve the equation $4x = 9 \pmod{11}$.

$$\begin{aligned} 4x &= 9 \quad \pmod{11} \\ 3(4x) &= 3(9) \pmod{11} \\ x &= 5 \quad \pmod{11} \end{aligned}$$

"So if you make 5 moves of $< 4 >$ on an 11 board, you land on [9]. However, $< 4 >$ makes a complete orbit every eleven moves.

Answers, 7.5

1. No, since $\gcd(15, 18) \nmid 5$.
2. Yes, $\gcd(8, 10) \mid 6$.
3. Yes, $\gcd(5, 13) \mid 3$.

Thus in 5 + 11 or 16 moves of <4> you will also land on the square [9], and again in 27 moves. Indeed, for any whole number t, you will land on [9] if you make $5 + 11t$ moves of < 4 >.

"In general, if it is possible to reach a square [k] on an m board, then it is possible to do so in fewer than m moves.* That means if [k] is in the orbit of < r > on an m board, there is some x less than m—in other words, some x in $\mathbf{Z}_m$—such that $rx = k$ (mod m)."

Exercise 7.6

For each of the following find the smallest whole number x, as well as a description of *all* whole numbers x, such that:

1. x moves of < 3 > will land on [4] on an 8 board;
2. x moves of < 6 > will land on [11] on a 13 board;
3. x moves of < 5 > will land on [3] on a 29 board;
4. x moves of < 13 > will land on [24] on a 40 board;
5. x moves of < 12 > will land on [14] on a 17 board;
6. x moves of < 12 > will land on [14] on a 15 board;
7. x moves of < 9 > will land on [6] on a 21 board.

"I know we solved equations like $9x = 6$ (mod 21) several sections ago, Ant. But all I remember is that they were a real mess. Let's see. I know that [6] is in the orbit of < 9 > on a 21 board since gcd(9, 21) = 3 and 3 is a factor of 6. (I also know that it is possible to reach [6] in fewer than 21/gcd(21, 9) = 7 moves.) But

---

*Every m times a man moves < r >, he lands back on [0] since the size of the orbit of < r > is a factor of m. (Alternately, m moves of < r > has the same effect as r moves of < m >, and a move of < m > on an m board returns you to [0].) Since each time a man returns to [0] the cycle begins anew, any square in the orbit of < r > on an m board can be reached in less than m moves.

Answers, 7.6

1. $3x = 4$ (mod 8) so $x = 4$. The size of the orbit of <3> on an 8 board is 8 So, for each whole number t, if you move <3> $4 + 8t$ times you also land on [4].
2. $6x = 11$ (mod 13) so $x = 4$. The size of the orbit is 13. So, for each whole number t, if you move < 6 > $4 + 13t$ times you also land on [11].
3. $5x = 3$ (mod 29) so $x = 18$. The size of the orbit is 29. So, for each whole number t, if you move < 5 > $18 + 29t$ times you will land on [3].
4. $13x = 24$ (mod 40) so $x = 8$. The size of the orbit is 40. So, for each whole number t, if you move < 13 > $8 + 40t$ times you will land on [24].
5. $12x = 14$ (mod 17) so $x = 4$. The size of the orbit is 17. So, for each whole number t, if you move < 12 > $4 + 17t$ times you will land on [14].
6. There is no x such that $12x = 14$ (mod 15) since gcd(12,15) $\nmid$ 14.
7. See the discussion of $9x = 6$ (mod 21) following the exercise.

to solve the equation we're supposed to go through some routine which, as I distinctly recall, you said was tedious to apply. Maybe that's why I like prime mods better."

"Parts 1 and 4 involved composite mods, Gnam, and you solved them. The first five parts were all fairly straightforward, because the equations were of the form $ax = b \pmod{m}$ where $/a \pmod{m}$ exists. But in parts 6 and 7, $/a \pmod{m}$ doesn't exist, because $\gcd(a, m) > 1$. In part 6 the square isn't in the orbit of the move (the equation has no solutions). But as you have noted, the equation in part 7 does have solutions.

"To find x in Z_{21} such that $9x = 6 \pmod{21}$ you could use one of the strategies we developed back in Section 4. But since you seem to have forgotten them, here's another way to reason: if $9x = 6 \pmod{21}$ then $21 \mid (9x - 6)$. Thus there exists y such that $21y = 9x - 6$, or, $9x - 21y = 6$. This is just the Football Problem, Gnam. Divide both sides of the equation by 3, obtaining $3x - 7y = 2$. Thus, $3r(x) = 2 \pmod{7}$ or $r(x) = 3 \pmod{7}$ where x is an element of $\mathbf{Z}_{21}$.* Hence the solutions of $9x = 6 \pmod{21}$ are 3, 10, and 17. 3 is the smallest whole number solution. Now, the orbit size of $< 9 >$ on a 21 board is $21/\gcd(9, 21) = 7$. So the general form for the number of moves of $< 9 >$ that land on [6] on a 21 board is $3 + 7t$."

"I knew this book would degenerate into nonsense!"

"Okay, hot shot, then *you* explain why you can find the smallest whole number x such that $9x = 6 \pmod{21}$ by solving the equation $3r(x) = 2 \pmod{7}$. I'll give you a few minutes to work on it. Meanwhile, our readers, who I feel confident followed every word of my explanation, can try their hand at the following problem."

Supplementary Problem 7.1

> In order to find the minimum number of moves of $< 9 >$ it takes to reach [6] on a 21 board, we needed to solve $9x = 6 \pmod{21}$. But this led to the Diophantine equation $3x - 7y = 2$. When $x = 3$, $y = 1$. 3 is the minimum number of moves required. Interpret the meaning of y. When $x = 10$, $y = 4$. What does this y mean? $(-4, -2)$ is another solution to the equation. Interpret its meaning.

---

*Alternately, you can reason: if $9x = 6 \pmod{21}$ then $21 \mid (9x - 6)$, so $7 \mid (3x - 2)$. Thus, $r(3x - 2) = 0 \pmod{7}$. So $3r(x) = 2 \pmod{7}$.

"I knew you were making this whole thing unnecessarily complicated! It's so typical of you, Ant. Here's an easy way to understand: On a board which is 21 units long, the squares which are in the orbit of < 9 > are the multiples of 3. So, the only squares of any concern are the squares [0], [3], [6], [9], [12], [15], and [18]. Hence, I can ignore all the other squares and picture the situation as follows:

[0]	[3]	[6]	[9]	[12]	[15]	[18]	21 board
[0]	[1]	[2]	[3]	[4]	[5]	[6]	7 board

"The move < 9 > on the 21 board corresponds to the move < 3 > on the 7 board. And getting to [6] on the 21 board is like getting to [2] on the 7 board. So instead of solving the equation

$$1.\ \ 9x = 6 \pmod{21}$$

(which is miserable since 9 has no mod 21 multiplicative inverse), solve instead the equation

$$2.\ \ 3x = 2 \pmod{7}.$$

"The point is that we could have gone from $9x = 6$ (mod 21) to $3x = 2$ (mod 7) by dividing everything in sight by 3, instead of going through the convoluted reasoning you went through with y's and such!"

"Equations 1 and 2 aren't equivalent, Gnam, because Equation 1 has *three* solutions, namely 3, 10, and 17, whereas Equation 2 has only one solution, namely 3. Still, you can use Equation 2 to find the solutions of Equation 1 so I suppose I shouldn't quibble."

"Right! You just solve equation 2 obtaining $x = 3$. So the first time you hit the square [6] on a 21 board is after moving < 9 > 3 times. Thereafter, every 7 times you move you hit [6] again. So, in general, you have to move $3 + 7t$ times! Pretty shrewd reasoning, don't you think? I believe I'll make up the next exercise, Ant."

Gnam's Exercise 1

Fill in the blanks:

1. The move < 6 > on a board 20 units long corresponds to the move < 3 > on a board ____ units long.
2. The move < 14 > on a board 35 units long corresponds to the move ____ on a board 5 units long.
3. The move < 15 > on a board 100 units long corresponds to the move < 3 > on a board ____ units long.
4. The move < 21 > on a board 70 units long corresponds to the move ____ on a board ____ units long.
5. The move < 21 > on a board 77 units long corresponds to the move ____ on a board ____ units long.
6. The move < 24 > on a board 42 units long corresponds to the move ____ on a board ____ units long.

Exercise 7.7

For each of the following find (a) the smallest whole number x and (b) a description of all whole numbers x such that:

1. x moves of < 6 > will land on [10] on a 20 board;
2. x moves of < 14 > will land on [21] on a 35 board;
3. x moves of < 15 > will land on [95] on a 100 board;
4. x moves of < 21 > will land on [28] on a 70 board;
5. x moves of < 21 > will land on [28] on a 77 board;
6. x moves of < 24 > will land on [36] on a 42 board;
7. x moves of < 10 > will land on [15] on an 18 board.

Answers, Gnam's Exercise 1

1. 10 2. < 2 > 3. 20

There are several correct answers for 4, 5, and 6, but only one where $\gcd(r,m) = 1$ namely:

4. < 3 >, 10 5. < 3 >, 11 6. < 4 >, 7

Answers, 7.7

1. $6x = 10 \pmod{20}$, so $3r(x) = 5 \pmod{10}$. $x = 5$; in general, x is of the form $5 + 10t$ where t is a whole number.
2. $14x = 21 \pmod{35}$ so $2r(x) = 3 \pmod 5$. $x = 4$; $4 + 5t$.
3. $15x = 95 \pmod{100}$ so $3r(x) = 19 \pmod{20}$. $x = 13$; $13 + 20t$.
4. $21x = 28 \pmod{70}$ so $3r(x) = 4 \pmod{10}$. $x = 8$; $8 + 10t$.
5. $21x = 28 \pmod{77}$ so $3r(x) = 4 \pmod{11}$. $x = 5$; $5 + 11t$.
6. $24x = 36 \pmod{42}$ so $4r(x) = 6 \pmod 7$. $x = 5$; $5 + 7t$
7. $10x = 15 \pmod{18}$ has no solutions since $\gcd(10,18) \nmid 15$. [15] is not in the orbit of < 10 > on an 18 board.

Exercise 7.8

Suppose that the starting square is not [0], but [7].

1. 4 moves of $< r >$ lands you on [6] on an 8 board. What is r?
2. x moves of $< 3 >$ initially lands you on [2] on a 13 board. What is x?
3. 3 moves of $< 6 >$ lands you on [7] on an m board. What is m?

Exercise 7.9

Which square do you start on if:

1. you land on [3] after 5 moves of $< 7 >$ on a 14 board?
2. you land on [8] after 6 moves of $< 3 >$ on a 37 board?

Supplementary Problem 7.2

Suppose that $\gcd(a,n) \mid b$, where a and b are elements of $\mathbf{Z}_n$. How many elements x of $\mathbf{Z}_n$ satisfy $ax = b \pmod{n}$? Suppose that x_0 is one solution of this equation. Express all the solutions in terms of x_0.

"You've dallied around long enough, Ant. Let's get to the chessboard!"

"Very well. We will now consider the orbits of men which move r squares to the right and u squares up on an m by n board."

Notation

$<r, u>$ represents the move r squares to the right and u squares up.

"Note that the move <8, 7> on a 5 by 3 board is equivalent to the move <3, 1> on this board. In general, we will only consider moves $<r, u>$ on an m by n board where $0 \le r < m$ and $0 \le u < n$.

Answers, 7.8

1. $7 + 4r = 6 \pmod 8$ has no solutions.
2. $7 + 3x = 2 \pmod{13}$ so $x = 7$. In general, x is of the form $7 + 13t$ where t is a whole number.
3. $7 + 3(6) = 7 \pmod m$, so $m = 18$ or $m = 9$ (the only factors of 18 which are larger than 7.)

Answers, 7.9

1. $x + 5(7) = 3 \pmod{14}$ so $x = 10$.
2. $x + 6(3) = 8 \pmod{37}$ so $x = 27$.

Also, unless stated otherwise, we will assume that all men begin on the square [0, 0].

"Suppose that you make two moves of <8, 5> on an 11 by 10 board. You will land on the square $[k, j]$ where:

$$2(8) = k \pmod{11}$$

and

$$2(5) = j \pmod{10}.$$

"Thus you will land on the square [5,0]."

Exercise 7.10

Suppose that a man is on the square [2, 3] on a 5 by 7 board.

1. On what square will he land on if he makes one move of:
   a. <2, 3>? b. <4, 1>? c. <3, 6>?
2. Answer 1 if he moves five times.

Exercise 7.11

1. Starting at [0, 0], a man moves <r, u> twice on a 15 by 14 board and lands on [6, 6]. What are the possible values for <r, u>?
2. Starting at [3, 2], a man moves <r, u> 5 times on a 6 by 9 board and lands on [0, 1]. What are the possible values for <r, u>?
3. On an 11 by 7 board, a man moves <5, 6> 10 times, and lands on [3, 4]. On which square did the man start?
4. On an 11 by 13 board, a man moves <3, 4> 15 times and lands on [0, 0]. On which square did the man start?
5. Starting at [3, 5], a man moves <2, 2> 15 times and lands on [4, 7]. What are the possible board sizes?
6. Starting at [0, 1], a man moves <3, 8> 10 times and lands on [3, 5]. What are the possible board sizes?

Answers, 7.10

1.	a.	[4, 6]	b.	[1, 4]	c.	[0, 2]
2.	a.	[2, 4]	b.	[2, 1]	c.	[2, 5]

Answers, 7.11

1. <3, 3> or <3, 10>; $2r = 6 \pmod{15}$ and $2u = 6 \pmod{14}$. So $r = 3$ and $u = 3$ or $u = 10$.
2. <3, 7>; $3 + 5r = 0 \pmod 6$ and $2 + 5u = 1 \pmod 9$. So, $r = 3$ and $u = 7$.
3. [8, 0]. $k + 10(5) = 3 \pmod{11}$ and $j + r_7(10) \cdot 6 = 4 \pmod 7$. $(k, j) = [8, 0]$.
4. [10, 5]. $k + r(15) \cdot 3 = 0 \pmod{11}$ and $j + r(15) \cdot 4 = 0 \pmod{13}$. $[k, j] = [10, 5]$.
5. The possible board sizes are 29 by 28 and 29 by 14.
6. There are several possible m by n boards. m must be a factor of 27 which is greater than 3; n must be a factor of 76 which is greater than 5. For example, the board could be 9 by 19.

"Once again, you have demonstrated your highly polished technique of wandering far off the point. Let me repeat my question of several pages ago, Ant. How can I tell whether a particular square is in the orbit of one of my men, and, if it is, how many times must I move the man to reach the square?"

"Gnam, you need to learn how to savor a problem. Turn it this way and then that way. Play with variations of the theme. That is how one acquires a sense of 'the lay of the land'. You might as well be a horse with blinders, if you're going to pay attention only to the carrot! But, very well! Let's begin investigating orbits. Recall that the orbit of a move is just the set of squares that can be reached by a man with that move, if he starts on [0, 0]. So, for example, if he moves <2, 2> on a 4 by 3 board, he successively reaches the squares [2, 2], [0, 1], [2, 0], [0, 2], [2, 1] and then lands back on [0, 0] to begin the cycle anew. The set of these six squares is the orbit of <2, 2> on a 4 by 3 board."

Exercise 7.12

Determine the orbits of each of the following moves on the indicated board.

1. <2, 3> on a 4 by 6 board
2. <2, 3> on a 4 by 8 board
3. <2, 3> on a 5 by 4 board
4. <4, 6> on a 8 by 18 board
5. <3, 2> on a 4 by 8 board

"Recall that the size of the orbit of a move is just the number of squares in the orbit. So, for example, the orbit sizes of the moves in the previous exercise are 2, 8, 20, 6, and 4, respectively. To determine the size of an orbit, it isn't necessary to list all of the squares in the orbit. To find the orbit size of, say, <6, 8> on a 10 by 16 board, you can reason as follows:

Horizontally: Every 5 times you move <6, anything> on a 10 by (Whatever) board you return to the same column you started in, since 10/gcd(6, 10) = 5.

Answers, 7.12

1. {[0, 0], [2, 3]};
2. {[0, 0], [2, 3], [0, 6], [2, 1], [0, 4], [2, 7], [0, 2], [2, 5]};
3. {[0, 0], [2, 3], [4, 2], [1, 1], [3, 0], [0, 3], [2, 2], [4, 1], [1, 0], [3, 3], [0, 2], [2, 1], [4, 0], [1, 3], [3, 2], [0, 1], [2, 0], [4, 3], [1, 2], [3, 1]} (The entire board);
4. {[0, 0], [4, 6], [0, 12], [4, 0], [0, 6], [4, 12]};
5. {[0, 0], [3, 2], [2, 4], [1, 6]}.

Vertically: Every 2 times you move <anything, 8> on a (Whatever) by 16 board you return to the same row you started in, since $16/\gcd(8, 16) = 2$.

Thus, every 10 times you move <6, 8> on a 10 by 16 board, you return to the same square you started on, and begin the cycle over again. (This also happens every 20 times, or 30 times, or, in general, $10k$ times you move. But it takes 10 moves to initially return you to your starting square.) So, the size of the orbit is 10."

Exercise 7.13

1. For each of the following moves, find the size of its orbit on the indicated board.
   a. <2, 4> on a 6 by 8 board;
   b. <2, 3> on a 6 by 4 board;
   c. <1, 5> on a 3 by 30 board;
   d. <4, 6> on a 14 by 20 board;
   e. <3, 4> on a 5 by 5 board.
2. Continue considering specific cases until you can answer: What is the size of the orbit of $<r, u>$ on an m by n board?

Theorem 7.3

The size of the orbit of $<r, u>$ on an m by n board is $\text{lcm}[m/\gcd(r, m), n/\gcd(u, n)]$.

"I know the next question, Ant: Which squares $[k, j]$ are in the orbit of the move $<r, u>$ on an m by n board? I even know its answer!"

"If so, I'm very impressed."

"It's nothing, Ant. I just generalized from the one dimensional case. The answer is: $[k, j]$ is in the orbit of $<r, u>$ on an m by n board if and only if $\gcd(r, m) \mid k$ *and* $\gcd(u, n) \mid j$."

"The conditions you state are *necessary* for a square to be in the orbit, that is, if they aren't met the square has no chance! But they aren't *sufficient*. For example, look at part 5 of Exercise 7.12. The square [3, 4] isn't in the orbit of <3, 2> on a 4 by 8 board. However, $\gcd(3, 4) \mid 3$ and $\gcd(2, 8) \mid 4$. Finding both necessary

---

Answers, 7.13

1. a. 6 b. 12 c. 6 d. 70 e. 5
2. See Theorem 7.3.

and sufficient conditions for a square to be in the orbit of a move on a specific board is a nontrivial problem. (I am going to pose it as a "Challenger" later in this section.)

"Meanwhile, I suggest we proceed under the (sometimes false) assumption that the squares in question are in the orbits of the moves under consideration and try to deduce how many times one must make the move to reach the desired square. If the square isn't in the orbit of the move, we'll find that out when our equations have no solutions.

"Thus, anytime I ask 'how many times must you move whatever to reach the square [wherever] on a (this) by (that) board', the answer may well be 'sorry, no soap!' That is, despite the assertive way I pose the question, the square may not be in the orbit of the move.

"So, let's consider now the question of how many moves of <2, 3> it takes to reach the square [3, 0] on a 5 by 4 board."

"That's easy. The answer is 4, Ant."

"How did you get that so quickly?"

"I just looked at the answer to part 3 of Exercise 7.12."

"Very well, let's change the question to how many moves of <2, 3> it takes to reach the square [8, 2] on a 9 by 7 board. If you land on [8, 2] in x moves of <2, 3> then:

$$2r_9(x) = 8 \text{ (mod 9) and } 3r_7(x) = 2 \text{ (mod 7)."}$$

"Hold it, Ant! What's with the r stuff? Why not just write x as we did before?"

"Because the smallest whole number x that works might be larger than 9, Gnam. And then it wouldn't be an element of either $\mathbf{Z}_7$ or $\mathbf{Z}_9$ and the above equations wouldn't make sense. As a matter of fact, I just figured out that the smallest x is 31. And we can't very well check it by saying:

$$2(31) = 8 \text{ (mod 9) and } 3(31) = 2 \text{ (mod 7).}$$

since 31 isn't an element of either of the algebraic structures under question. If I were to write such a thing, it would be considered blatantly incorrect by those who believe in using the language correctly, and sloppy by those with a more charitable nature."

"I'm very charitable, Ant. And I don't object to being sloppy when it saves me from having to deal with confusing notation. I can recall your grumbling over the sloppy notation used in calculus texts, and yet you continue to use it because it works."

"Okay, Gnam. But if anyone complains, I'm blaming you."

"You were saying that to find the number of times you have to move <2, 3> to reach [8, 2] on a 9 by 7 board, just solve

$$2x = 8 \pmod 9 \text{ and } 3x = 2 \pmod 7.$$

"But how do you do that, Ant? If you solve for x in the first equation you get 4, and if you solve for x in the second equation you get 3. Those are different x's! How did you get 31 from 4 and 3?"

"That's what happens when you use sloppy notation! You *do* get confused. What we really figured out was that $r_9(x) = 4$ and $r_7(x) = 3$. Therefore x must be 4 more than a multiple of 9, and x must also be 3 more than a multiple of 7."

"Oh, I really knew that. Keep it up with the sloppy notation."

"$x = 4 + 9t$ and $x = 3 + 7s$ for some whole numbers s and t. Thus,

$$3+7s=4+9t$$
$$7s-9t=1.$$

"Now, solving this equation using the technique we developed in the last section we get:

$$-2r(t)=1 \pmod 7$$
$$r(t)=3 \pmod 7)*.$$

"Therefore t is of the form $3 + 7v$. Substituting this into the equation $x = 4 + 9t$, we obtain $x = 4 + 9(3 + 7v) = 31 + 63v$. Clearly 31 is the smallest whole number value for x. (Let $v = 0$.) But for any whole number v, if you make $31 + 63v$ moves of <2, 3> on a 9 by 7 board, you will reach [8, 2]. Letting $v = 1$, we obtain $x = 94$. So in 94 moves of <2, 3> you also reach [8, 2]. Indeed, every 63 moves, <2, 3> makes a complete orbit.

"Since we already know a different way to find the size of an orbit, let's focus on finding the *minimum* number of times we must move to reach a desired square."

---

*What does $t = 3$ mean in the context of this problem?

"Hold it, Ant! Why didn't you also solve for s in the equation $7s - 9t = 1$? I thought that when you solved an equation, you had to solve for everything!"

'You have to learn to keep your eyes on the carrot, Gnam. All we care about is finding x, and solving for either t or s will give us that. Here, look at what happens when you solve for s:

$$7s - 9t = 1$$
$$7r(s) = 1 \pmod 9$$
$$r(s) = 4 \pmod 9.$$

"Substituting $4 + 9v$ for s, we obtain $x = 3 + 7s = 3 + 7(4 + 9v) = 31 + 63v$. Look familiar? The point is that in order to find x, it is necessary only to solve for either s or t, not *both*."

"Let's do some more examples, Ant."

Example

What is the minimum number of times you must move <7, 2> to reach [6, 4] on a 29 by 31 board?

We seek the smallest whole number x such that:

$$7x = 6 \pmod{29} \quad \text{and} \quad 2x = 4 \pmod{31}.$$

So:

$$x = 5 \pmod{29} \quad \text{and} \quad x = 2 \pmod{31}.$$

Thus $x = 5 + 29t$* and $x = 2 + 31s$.

So $$5 + 29t = 2 + 31s$$

or $$31s - 29t = 3$$
$$2r(t) = 3 \pmod{31}$$
$$r(t) = 17 \pmod{31}.$$

The smallest whole number satisfying $r(t) = 17 \pmod{31}$ is clearly 17. Substituting 17 for t, we have $x = 5 + 29t = 5 + 29(17) = 498$. So you need to move <7, 2> 498 times before you reach the square [6, 4] on a 29 by 31 board!

---

*At this point, you can save a step by substituting $5 + 29t$ for x in the equation $x = 2 \pmod{31}$, obtaining $5 + 29t = 2 \pmod{31}$. Solving, you find that the smallest whole number t which satisfies this equation is 17. So $x = 5 + (29)(17) = 498$.

Example

What is the minimum number of times you must move <6, 21> to reach [3, 28] on a 15 by 77 board?

We seek the smallest whole number x such that

$$6x = 3 \pmod{15} \quad \text{and} \quad 21x = 28 \pmod{77}.$$

So $\quad 2x = 1 \pmod 5 \quad$ and $\quad 3x = 4 \pmod{11}$.

Thus $\quad x = 3 + 5t \quad$ and $\quad x = 5 + 11s$.

$$3 + 5t = 5 + 11s$$

$$5t - 11s = 2$$

Solving for s we get $r(s) = 3 \pmod 5$. So the smallest x is $x = 5 + 11(3) = 38$.

Example

What is the minimum number of times you must move <3, 7> to land on the square [5, 2] on a 12 by 8 board?

We seek the smallest whole number x such that

$$3x = 5 \pmod{12} \quad \text{and} \quad 7x = 2 \pmod 8.$$

But the equation $3x = 5 \pmod{12}$ has no solutions, so [5, 2] isn't in the orbit of <3, 7> on a 12 by 8 board.

Example

What is the minimum number of times you must move <13, 49> to land on [7, 0] on a 120 by 98 board?

We seek the smallest whole number x such that

$$13x = 7 \pmod{120} \quad \text{and} \quad 49x = 0 \pmod{98}$$

or $\quad 13x = 7 \pmod{120} \quad$ and $\quad x = 0 \pmod 2$.

Solving the first equation, we get

$$\begin{aligned} 13x &= 7 && \pmod{120} \\ 37(13x) &= 37(7) && \pmod{120} \\ x &= 19 && \pmod{120}. \end{aligned}$$

So $x = 19 \pmod{120}$ and $x = 0 \pmod 2$.

(It may be evident to you that no x will work, since no even number leaves a mod 120 remainder of 19. But let's pretend that we didn't notice, and continue solving.)

$$19 + 120t = 0 + 2s$$

$$2s - 120t = 19$$

This equation has no solutions since gcd(2, 120) is not a factor of 19. Hence, the square [7, 0] is not in the orbit of <13,49> on a 120 by 98 board.

Exercise 7.14

In each of the following, you are given an m by n board, a move <r, u>, and a square [k, j]. Determine the minimum number of times you must move to land on the square.

	size of board	move	ending square
1.	7 by 11	<5, 10>	[2, 3]
2.	19 by 31	<7, 11>	[2, 3]
3.	12 by 21	<8, 14>	[4, 7]
4.	11 by 9	<5, 6>	[8, 2]

Exercise 7.15

In each of the following, you are given an m by n board, a move <r, u>, and a square [k, j]. Assume that the starting square is [2, 3]. Determine in general the number of times you must move to land on the square.

	size of board	move	ending square
1.	5 by 12	<3, 6>	[2, 0]
2.	6 by 9	<2, 5>	[0, 6]
3.	19 by 24	<5, 11>	[16, 12]
4.	18 by 30	<5, 11>	[16, 23]
5.	12 by 15	<10, 10>	[10, 8]
6.	18 by 24	<3, 9>	[8, 15]

Answers, 7.14

1. 41 2. 516 3. 2
4. The square isn't in the orbit of the move on this board since gcd(6, 9) $\nmid$ 2

Answers, 7.15

1. The corresponding equations are $2 + 3x = 2 \pmod 5$ and $3 + 6x = 0 \pmod{12}$. The latter equation, which is equivalent to $6x = 9 \pmod{12}$, has no solutions since gcd(6, 12) $\nmid$ 9. Hence, the square [2, 0] can't be reached.
2. The square can't be reached since $6 + 9t = 2 + 3s$ has no solutions, for gcd(3, 9) $\nmid$ 4.
3. $\{75 + 456v$: v is a whole number$\}$
4. $\{10 + 90v$: v is a whole number$\}$
5. $\{2 + 6v$: v is a whole number$\}$
6. $\{20 + 24v$: v is a whole number$\}$

Exercise 7.16

What are the possible orbit sizes for moves on a

1. 7 by 5 board?
2. 7 by 8 board?
3. 12 by 8 board?
4. m by n board?

Exercise 7.17

For each possible orbit size on a 24 by 16 board, find a move with that orbit size.

Supplementary Problem 7.3

On the 5 by 11 by 18 *three*-dimensional board, what is the minimum number of times you must make the move <2, 5, 1> to land on the square [1, 2, 15]?

Suggested strategy: To simultaneously solve three equations, first simultaneously solve any two of them. Then simultaneously solve that solution (which is itself an equation) with the third equation.

Challenger

Determine necessary and sufficient conditions for the square $[k, j]$ to be in the orbit of the move <r, u> on an m by n board.

Answers, 7.16

1. 1, 5, 7, and 35
2. 1, 2, 4, 7, 8, 14, 28, and 56
3. 1, 2, 3, 4, 6, 8, 12, and 24
4. The factors of lcm(m, n)

Answer, 7.17

The possible orbit sizes are the factors of lcm(24, 16) = 48, namely 1, 2, 3, 4, 6, 8, 12, 16, 24, and 48. The only move with orbit size 1 is <0, 0>. There are several moves corresponding to each of the other orbit sizes. One set of possible answers follows. 2: <12, 8>; 3: <16, 0>; 4: <6, 8>; 6: <8, 8>; 8: <3, 0>; 12: <10, 8>; 16: <0, 5>; 24: <1, 0>, and 48: <5, 3>.

Section Eight

Miscellaneous Problems

We have to go now. As a parting gift, we offer you a collection of number theory problems which have been gathered from assorted sources, mostly *The Mathematics Teacher*, *Consortium*, and our imaginations.

Ant, **Gnam**, and *Andrea*

1. Long, long ago Brad needed to buy 100 head of livestock with exactly \$100. Cows were \$10 each, hogs were \$3 each, and chickens were 50 cents each. If Brad had to buy at least one of each kind of animal, how many of each did he buy?

2. Show that there is no integer n such that $4 \mid (7n - 3)$ and $6 \mid (5n + 2)$.

3. What number less than 200
   leaves a remainder of 1 when divided by 2 and
   leaves a remainder of 2 when divided by 3 and
   leaves a remainder of 3 when divided by 4 and
   leaves a remainder of 4 when divided by 5 and
   leaves a remainder of 5 when divided by 6 but
   leaves no remainder when divided by 7?

4. If four days before tomorrow is Thursday, what is three days after yesterday?

5. What is the smallest three-digit number which is divisible by the first three positive primes and by the first three positive composites?

6. What is the smallest positive integer which is divisible by 2, 3, 7, 10, 15, 20, 21, and 28?

7. What is the least number of pennies that can either be arranged in 9 equal piles, 10 equal piles, 11 equal piles, or 12 equal piles?

8. Deon, a conductor, wants to arrange the members of her band first into a perfect square. Then she wants to rearrange them by dividing all but the 3 drum players into 4 rectangles of equal size. Is Deon also a number theorist? How do you know?

9. Lori collects seashells. If she counts the shells by twos, threes, fours, fives, or sixes, one is always left over. However, counting them by sevens comes out even. What is the fewest number of seashells that Lori could have?

10. Jamie is a slave to his schedules. (For example, he buys gas every 19 days whether his car needs it or not!) He is similarly fastidious about his three household responsibilities: mopping, dusting, and vacuuming. One Wednesday, (in a non-leap year), Jamie did all three of these chores. Vacuuming being his least favorite task, he noted with relief that he wasn't scheduled to vacuum again until the Tuesday of the week after next, by which time he would have dusted another three times and mopped floors once (but not on a day he was scheduled to dust.) Also, when he did have to vacuum again, at least he would have just enjoyed a long weekend (Friday through Sunday) free of chores.

    There is only one other day in the same year when Jamie will again perform all three chores. What is its date?

11. Jamie went grocery shopping (in a state with no sales tax) and bought bread (at 90 cents per loaf), soup (at \$1.25 per can), gum (at 65 cents per pack), and nothing else. In a rush to get home to do his household chores, he bought only 10 items so that he could use the express lane. He handed the cashier a \$10 bill, and the cashier gave him a \$1 bill and two like coins in change. How many packs of chewing gum did Jamie buy?

12. Prove that if the integers $n - 10$, $n - 6$, $n + 6$, and $n + 10$ are all prime numbers, then so is n.

13. Adjoin to the right of the digits 3456 three more digits so that the resulting number is divisible by 7, 8, and 11.

14. Find a whole number n such that $1 + 2 + 3 + ... + n$ is a three-digit number whose digits are all the same. (Note: $1 + 2 + 3 + ... + n = n(n+1)/2$. One way to solve this problem is by using a mod 37 argument.)

15. The number 33 is a multiple of 11 and leaves a remainder of 1 upon division by 4. Find the next largest positive integer with both of these properties. Find the largest negative integer with both of these properties.

16. Prove that for all nonnegative integers n,

$$5 \mid 2^{2n+1} + 3^{2n+1}, \qquad 17 \mid 2^{6n+3} + 3^{4n+2},$$
$$7 \mid 2^{n+2} + 3^{2n+1}, \qquad 19 \mid 2^{3n+4} + 3^{3n+1},$$
$$11 \mid 2^{8n+3} + 3^{n+1}, \qquad 29 \mid 2^{5n+1} + 3^{n+3},$$
$$13 \mid 2^{4n+2} + 3^{n+2}, \quad \text{and} \quad 31 \mid 2^{4n+1} + 3^{6n+9}.$$

17. Ms. Bruce, a teacher, writes a positive integer less than fifty thousand on the board. One student claims it is a multiple of 2; a second student claims it is a multiple of 3; and so on, until the twelfth student claims it is a multiple of 13. Ms. Bruce observes that all but two of the students are right, and that the two students who made incorrect statements spoke consecutively. What is the number written on the board?

18. Which nine-digit number, containing one each of the digits 1 through 9 in some order, satisfies all of the following? The number is divisible by 9. If you cross out the rightmost digit, the remaining eight-digit number is divisible by 8. If you cross out the rightmost two digits, the remaining seven-digit number is divisible by 7, *etc.* That is, for each n from 0 to 8, if you cross out the rightmost n digits, the remaining $9 - n$ digit number is divisible by $9 - n$.

Section Nine

Solutions to Problems

Supplementary Problem 2.1

One way to extend the Division Algorithm so that it applies to all integral divisors, other than 0, is as follows. If a and n are integers, $n \neq 0$, then there exist unique integers q and r such that $a = qn + r$ where $0 \le r < |n|$. So, for example, if $a = 27$ and $n = -4$, then $q = -6$ and $r = 3$ since $27 = (-6)(-4) + 3$ and $0 \le 3 < |-4|$. Similarly, if $a = -27$ and $n = -4$, then $q = 7$ and $r = 1$ since $-27 = 7(-4) + 1$ and $0 \le 1 < |-4|$.

Supplementary Problem 2.2

1. By the Division Algorithm, there exist unique q and r such that $b = 11q + r$ where $0 \le r < 11$. So for any integer a, $11a + b = 11a + (11q + r) = 11(a + q) + r$. So when $11a + b$ is divided by 11, the remainder is r (and the quotient is $a + q$). That is, $r_{11}(11a + b) = r = r_{11}(b)$.
2. By the Division Algorithm, there exist unique q_1 and r_1 such that $s = 11q_1 + r_1$, $0 \le r_1 < 11$, and there exist unique q_2 and r_2 such that $t = 11q_2 + r_2$, $0 \le r_2 < 11$.

   So, $s + t = (11q_1 + r_1) + (11q_2 + r_2) = 11(q_1 + q_2) + (r_1 + r_2)$.

   $$\begin{aligned} \text{Hence: } r_{11}(s+t) &= r_{11}[11(q_1+q_2)+(r_1+r_2)] \\ &= r_{11}(r_1+r_2) && \text{by part 1} \\ &= r_1 +_{11} r_2 && \text{by definition of } +_{11} \\ &= r_{11}(s) +_{11} r_{11}(t). \end{aligned}$$

   Similarly,
   $s \cdot t = (11q_1 + r_1) \cdot (11q_2 + r_2) = 11[11q_1q_2 + q_1r_2 + q_2r_1] + r_1 \cdot r_2$.

   $$\begin{aligned} \text{Hence: } r_{11}(s \cdot t) &= r_{11}(r_1 \cdot r_2) && \text{by part 1} \\ &= r_1 \cdot_{11} r_2 && \text{by definition of } \cdot_{11} \\ &= r_{11}(s) \cdot_{11} r_{11}(t). \end{aligned}$$

Supplementary Problem 2.3

1. On one hand, $0 + k = 0$. But on the other hand, $0 + k = k$. Hence, $0 = k$. (This shows that a ring has precisely one additive identity.)

2. On one hand, $d + (x + d') = d + 0 = d$. But on the other hand, $d + (x + d') = (d + x) + d' = 0 + d' = d'$. Hence, $d = d'$. (This shows that an element of a ring can have at most one additive inverse.)

3. If $1k = 1$ and $1k = k$ it follows that $k = 1$. (This shows that a ring has precisely one multiplicative identity.)

4.
$$\begin{aligned} 0+a\cdot 0 &= a\cdot 0 && \text{identity property of } 0\\ &= a(0+0) && \text{identity property of } 0\\ &= a\cdot 0+a\cdot 0. && \text{distributive property}\end{aligned}$$

so $\quad 0+a\cdot 0=a\cdot 0+a\cdot 0.$

Thus,

$$\begin{aligned} (0+a\cdot 0)+-(a\cdot 0) &= (a\cdot 0+a\cdot 0)+-(a\cdot 0) \\ 0+[a\cdot 0+-(a\cdot 0)] &= a\cdot 0+[a\cdot 0+-(a\cdot 0)] && \text{assoc. prop. of addition}\\ 0+0 &= a\cdot 0+0 && \text{definition of additive inverse}\\ 0 &= a\cdot 0. && \text{identity property of } 0\end{aligned}$$

5.
$$\begin{aligned} -1a+a &= -1a+1a && \text{identity property of } 1\\ &= (-1+1)a && \text{distributive property}\\ &= 0\cdot a && \text{definition of additive inverse}\\ &= a\cdot 0 && \text{commutative property of mult.}\\ &= 0. && \text{part 4}\end{aligned}$$

Since $-1a + a = 0$, $-1a$ is the additive inverse of a. That is, $-1a = -a$.

6.
$$\begin{aligned} -a+-b &= -1a+-1b && \text{part 5}\\ &= -1(a+b) && \text{distributive property}\\ &= -(a+b). && \text{part 5}\end{aligned}$$

Alternately, one could argue:

$(-a+-b)+(a+b)=(-a+a)+(-b+b)$ repeated use of the comm. and assoc. properties of addition

$=0+0$ definition of additive inverse

$=0.$ identity property of 0

Since $(-a+-b)+(a+b)=0$, $-a+-b$ is the additive inverse of $a+b$. That is, $-a+-b=-(a+b)$.

7. $-a(b)=(-1a)b$ part 5

$=-1(ab)$ associative property of mult.

$=-(ab).$ part 5

By similar reasoning, $a(-b)=-(ab)$. Hence $(-a)b=-(ab)=a(-b)$.
Alternately, one could argue:

$(-a)b+ab=(-a+a)b$ distributive property

$=0b$ definition of additive inverse

$=b\cdot 0.$ commutative property of multiplication

$=0.$ part 4

So, $(-a)b=-(ab)$.

8. $a+(-a)=0$ so a is the additive inverse of $-a$. That is, $a=-(-a)$.

9. $(-a)(-b)=(-1a)(-1b)$ part 5

$=[(-1)(-1)]ab$ repeated use of the commutative and associative properties of mult.

$=[-(-1)]ab$ part 5

$=1ab$ part 8

$=ab.$ identity property of 1

10. Suppose that $a+b=a+c$.

Then $-a+(a+b)=-a+(a+c)$

$(-a+a)+b=(-a+a)+c$ assoc. property of addition

$0+b=0+c$ definition of additive inverse

$b=c.$ identity property of 0

Supplementary Problem 2.4

Mod 11 addition is commutative: For each a and b in $\mathbf{Z}_{11}$,

$$\begin{aligned} a +_{11} b &= r_{11}(a+b) && \text{definition of } +_{11} \\ &= r_{11}(b+a) && \text{since ordinary addition is commutative} \\ &= b +_{11} a. && \text{definition of } +_{11} \end{aligned}$$

Mod 11 addition is associative: For each a, b, and c in $\mathbf{Z}_{11}$,

$a+(b+c)=(a+b)+c$ since ordinary addition is associative.

So, $$r_{11}[a+(b+c)] = r_{11}[(a+b)+c].$$

By the Super - Important Theorem,

$$r_{11}(a) +_{11} r_{11}(b+c) = r_{11}(a+b) +_{11} r_{11}(c)$$

$$r_{11}(a) +_{11} [r_{11}(b) +_{11} r_{11}(c)] = [r_{11}(a) +_{11} r_{11}(b)] +_{11} r_{11}(c).$$

Since $a, b,$ and c are elements of $\mathbf{Z}_{11}$, $r_{11}(a) = a$, $r_{11}(b) = \mathrm{b}$, and $r_{11}(c) = c$.

So, $$a +_{11} (b +_{11} c) = (a +_{11} b) +_{11} c.$$

The other parts can be proved in a similar fashion.

Supplementary Problem 2.5

The equation $2x + 3y = 7$ over $\mathbf{Z}_{11}$ is equivalent to the equation $y = 3x + 6$ over $\mathbf{Z}_{11}$. Its solutions are (0, 6), (1, 9), (2, 1), (3, 4), (4, 7), (5, 10), (6, 2), (7, 5), (8, 8), (9, 0), and (10, 3). One way to picture the graph is as four "northeast lines" each with slope 3. Another way is to picture it as four "northwest lines" each with slope $-2/3$. Note that $-2/3 = -2 \cdot 4 = -8 = 3$ (mod 11).

Supplementary Problem 2.6

We have already observed that each nonzero element of $\mathbf{Z}_{11}$ has a unique multiplicative inverse, and hence $xy = 1$ if and only if $/x = y$ (mod 11).

1. $$\begin{aligned} (a/b)(b/a) &= [(a/b)b]/a && \text{associative property} \\ &= [a(/b \cdot b)]/a && \text{associative property} \\ &= (a \cdot 1)/a && \text{definition of mult. inverse} \\ &= a/a && \text{identity property of 1} \\ &= 1. && \text{definition of mult. inverse} \end{aligned}$$

2. $(a/b)(b/a) = 1$ by part 1 above. Hence, $/(a/b) = b/a$.

3. $(/a/b)(ab) = [(/a/b)a]b$ associative property

$= [a(/a/b)b$ commutative property

$= [(a/a)/b]b$ associative property

$= (1/b)/b$ definition of mult. inverse

$= /b \cdot b$ identity property of 1

$= 1.$ definition of mult. inverse

Thus, $/a/b = /(ab)$.

4. $(a/b)(c/d) = [(a/b)c]/d$ associative property

$= [c(a/b)]/d$ commutative property

$= [(ca)/b]/d$ associative property

$= (ca)(/b/d)$ associative property

$= (ca)/(bd)$ part 3

$= (ac)/(bd).$ commutative property

5. $a/b + c/d = (a/b)\cdot 1 + (c/d)\cdot 1$ identity property

$= (a/b)(d/d) + (c/d)(b/b)$ definition of mult. inverse

$= (ad)/(bd) + (cb)/(db)$ part 4

$= (ad)/(bd) + (bc)/(bd)$ commutative property

$= [ad + bc]/(bd)$ distributive property

6. Suppose $b \neq 0$ and $d \neq 0$. Then $a/b = c/d \Leftrightarrow (a/b)(bd) = (c/d)(bd) \Leftrightarrow ad = bc$. (Use arguments like those used above to obtain this last statement.)

7. $(a/b)/(c/d) = (a/b)(d/c)$ part 2

$= (ad)/(bc).$ part 4

8. $ac/bc = (a/b)(c/c)$ part 4

$= (a/b)\cdot 1$ definition of mult. inverse

$= a/b.$ identity property of 1

Supplementary Problem 2.7

1. $(x-8)(x-10)$
3. $(x-2)(x-9)$
4. $(x-8)(x-9)$
5. $3(x-3)(x-10)$

Supplementary Problem 3.1

The units are 1, –1, i, and $-i$. For each Gaussian integer $a + bi$, $1(a + bi) = a + bi$, so 1 is a unit; $-1(-a - bi) = a + bi$, so –1 is a unit;

$i(b - ai) = a + bi$, so i is a unit; and $-i(-b + ai) = a + bi$ so $-i$ is a unit. $2 = (1 + i)(1 - i)$ and $5 = (1 + 2i)(1 - 2i)$ so 2 and 5 are composite Gaussian integers.

If a prime integer, p, can be factored in a nontrivial way over the Gaussian integers, it can be expressed as $p = (a + bi)(a - bi)$. So $p = a^2 + b^2$. Thus, $r_4(p) = r_4(a^2) +_4 r_4(b^2) = [r_4(a)]^2 +_4 [r_4(b)]^2$. Now, for each x in $\mathbf{Z}_4$, $x^2 = 0$ or $x^2 = 1 \pmod 4$. Thus, $[r_4(a)]^2 +_4 [r_4(b)]^2$ is 0, 1, or 2. It can never be 3. So if $r_4(p) = 3$, then p is also a prime Gaussian integer.

It can also be shown that prime integers of the form $4k + 1$ can be expressed as the sum of two squares and thus are composite Gaussian integers. For example, $r_4(13) = 1$ and $13 = 9 + 4$ so $13 = (3 + 2i)(3 - 2i)$.

Supplementary Problem 3.2

Let $R[x]$ denote the set of all polynomials with real coefficients.

1. If b is a nonzero real number (that is, a polynomial of degree 0) and $f(x) = a_n x^n + a_{n-1}x^{n-1} + a_{n-2}x^{n-2} + \ldots + a_1 x + a_0$ is any element of $R[x]$, then $b \mid f(x)$ since
$$b\left(\frac{a_n x^n}{b} + \frac{a_{n-1}x^{n-1}}{b} + \frac{a_{n-2}x^{n-2}}{b} + \ldots + \frac{a_1 x^n}{b} + \frac{a_0}{b}\right) = f(x).$$
So b is a unit of $R[x]$.

2. The only way to factor $ax + b$, $a \neq 0$, is into a product of a unit and a polynomial of degree 1, so all polynomials of degree 1 are prime.

3. The polynomial $ax^2 + bx + c$ has real roots if and only if $b^2 - 4ac \geq 0$. The polynomial has roots r and s if and only if it can be factored into $a(x - r)(x - s)$. Thus, $ax^2 + bx + c$, where $a \neq 0$ is composite if $b^2 - 4ac \geq 0$, and prime if $b^2 - 4ac < 0$.

4. All polynomials of degree 3 have either 1 or 3 real roots and so can be factored over the reals into a product of a linear polynomial times a quadratic polynomial. (If the polynomial has three real roots, it can be factored into a product of three linear polynomials.) Hence all polynomials of degree 3 are composite.

5. All polynomials of degree 4 or larger are composite. For example, a polynomial of degree 4 can either have 4 (not necessarily distinct) real roots, (in which case one way to factor

the polynomial is into the product of 4 linear polynomials), 2 real roots and 2 complex roots (in which case one way to factor the polynomial is into the product of 2 linear and 1 quadratic polynomial), or 4 complex roots (in which case the polynomial can be factored into the product of 2 quadratic polynomials).

Supplementary Problem 3.3

1. If $a \mid b$ and $b \mid c$ then there exist integers s and t such that $as = b$ and $bt = c$. Thus, $a(st) = (as)t = bt = c$. So $a \mid c$.
2. If $a \mid b$ and $a \mid c$ then there exist integers s and t such that $as = b$ and $at = c$. Thus, $a(sx + ty) = (as)x + (at)y = bx + cy$. So $a \mid (bx + cy)$. Similarly, since $a(sx - ty) = bx - cy$, we know that $a \mid (bx - cy)$.
3. $c \neq 0$. Thus $a \mid b \Leftrightarrow$ there exists an integer s such that $as = b$ $\Leftrightarrow (ac)s = bc \Leftrightarrow ac \mid bc$.

Supplementary Problem 3.4

$(q_1+1)(q_2+1)(q_3+1)\cdots(q_k+1)$

Supplementary Problem 3.5

Suppose that a/b is a reduced fraction. Then $\gcd(a, b) = 1$. If a^3/b^3 is not a reduced fraction, then there exists a prime, p, such that $p \mid a^3$ and $p \mid b^3$. Now $a^3 = a^2a$ so by Theorem 3.2, $p \mid a^2$ or $p \mid a$. If $p \mid a^2$, then, again by Theorem 3.2, $p \mid a$. By similar reasoning, $p \mid b$. This contradicts the assumption that $\gcd(a, b) = 1$. Hence a^3/b^3 is a reduced fraction.

Supplementary Problem 3.6

The proofs are all nearly identical to the one given in the text that $\sqrt{2}$ is not rational. In part 5, note the assumption that a/b is a reduced fraction. To prove part 6, use the result of Supplementary Problem 3.5.

Supplementary Problem 4.1

Suppose that b is a nonzero element of $\mathbf{Z}_n$ and $\gcd(b, n) > 1$. Let $g = \gcd(b, n)$. So n/g and b/g are both integers. Now, n/g is also a nonzero element of $\mathbf{Z}_n$ since $g > 1$. Furthermore, $b(n/g) = n(b/g)$ which is a multiple of n. So $b(n/g) = 0 \pmod{n}$. Hence b is a mod n zero divisor.

Conversely, suppose that b is a zero divisor of $\mathbf{Z}_n$, that is, $bc = 0 \pmod{n}$ for some nonzero c in $\mathbf{Z}_n$. Then, $r_n(bc) = 0$. So, $n \mid bc$. We want to show that $\gcd(b, n) > 1$. Either $\gcd(b, n) > 1$ or $\gcd(b, n) = 1$.

Now, if $\gcd(b, n) = 1$, it follows from Theorem 3.3 that $n \mid c$. But c is an element of $\mathbf{Z}_n$ and so $0 \le c < n$. Therefore $c = 0$, which contradicts the assumption that b is a mod n zero divisor. Thus, $\gcd(b, n) > 1$.

Thus b is a mod n zero divisor if and only if $\gcd(b, n) > 1$.

Supplementary Problem 4.2

Suppose that b is a nonzero element of $\mathbf{Z}_n$ and $\gcd(b, n) > 1$. Then for any x in $\mathbf{Z}_n$, $bx = 0 \pmod{n}$ if and only if x is a multiple of $n/\gcd(b, n)$. Thus, letting $g = \gcd(b, n)$, x is $0, n/g, 2n/g, 3n/g, \ldots$ or $(g-1)n/g$. For example, if $n = 36$ and $b = 8$, then x must be a multiple of $36/\gcd(8, 36) = 36/4 = 9$. So x must be 0, 9, 18, or 27.

Supplementary Problem 4.3

If $ab = ba = 1$ and $ab' = b'a = 1$, then, on one hand $(b'a)b = 1b = b$. But, on the other hand, $(b'a)b = b'(ab) = b' \cdot 1 = b'$. Therefore, $b = b'$.

Supplementary Problem 4.4

Suppose that b is a mod n zero divisor. Then there exists a nonzero k in $\mathbf{Z}_n$ such that $bk = 0 \pmod{n}$. Now, if there were an element j of $\mathbf{Z}_n$ such that $bj = 1 \pmod{n}$, then, on one hand, $b(jk) = (bj)k = 1k = k$, and, on the other hand, $b(jk) = j(bk) = j(0) = 0$. Thus $k = 0$. But $k \neq 0$! So b cannot be a zero divisor and also have a multiplicative inverse.

Supplementary Problem 4.5

1. $a \cdot /a = 1$. Therefore, $a = /(/a)$.
2. $(ab)(/a \cdot /b) = (a/a)(b/b) = 1 \cdot 1 = 1$. So, $/a \cdot /b$ is the multiplicative inverse of ab. That is, $/a \cdot /b = /(ab)$.
3. $(-a)(-/a) = a/a = 1$. So $-/a$ is the multiplicative inverse of $-a$. That is, $-/a = /(-a)$. (You could also prove this using the result from Supplementary Problem 2.3: $-a = -1a$.)
4. If $ax = ay$ and a has a multiplicative inverse, then $/a(ax) = /a(ay)$. So $(/a \cdot a)x = (/a \cdot a)y$. So $x = y$.

Supplementary Problem 4.6

Suppose that b is a unit of an integral domain. Then $b \mid 1$. Hence b has a multiplicative inverse. Conversely, suppose that b has a multiplicative inverse, $/b$. Then for each element k of the integral domain, $b(/b \cdot k) = k$. So, $b \mid k$. Hence b is a unit. Characterizing an element as a unit is equivalent to stating that it has a multiplicative inverse.

Supplementary Problem 4.7

Suppose that the cancellation property of multiplication holds. If $ab = 0$, then $ab = a \cdot 0$. Hence, either $a = 0$, or we can cancel the a's and obtain $b = 0$. So, whenever $ab = 0$ it follows that $a = 0$ or $b = 0$. Thus the ring has no zero divisors. Conversely, suppose that the ring has no zero divisors. If $ab = ac$ and $a \neq 0$, it follows that $ab - ac = 0$, so $a(b - c) = 0$. Since $a \neq 0, b - c = 0$. So $b = c$. Thus, the cancellation property of multiplication holds. Therefore a commutative ring is an integral domain if and only if it has the cancellation property of multiplication.

Supplementary Problem 4.8

If $x = /x \pmod{p}$, then $x^2 = 1 \pmod{p}$. Hence, $x^2 - 1 = 0 \pmod{p}$. So, $(x - 1)(x + 1) = 0 \pmod{p}$. Now, $\mathbf{Z}_p$ has no zero divisors. So, $x - 1 = 0 \pmod{p}$ or $x + 1 = 0 \pmod{p}$. Hence $x = 1$ or $x = p - 1$.

Supplementary Problem 4.9

Since $b^2 = (-b)^2$, each perfect square, except 0, has at least two square roots. But, an element of a prime mod can't have more than two square roots. (Recall, in a prime mod if $x^2 = y^2$, then $x = y$ or $x = -y$.) Therefore, putting 0 aside, precisely half of the $p - 1$ remaining elements of $\mathbf{Z}_p$ are perfect squares. That is, there are $1 + (p - 1)/2$ mod p quadratic residues.

Supplementary Problem 4.10

If p is prime, each nonzero element of $\mathbf{Z}_p$ has a mod p multiplicative inverse. By Supplementary Problem 4.8, the elements 1 and $p - 1$ are the only two elements which are their own inverses. We could rewrite the factors of 10! (mod 11) as

$$\begin{aligned} 1 \cdot (2 \cdot 6) \cdot (3 \cdot 4) \cdot (5 \cdot 9) \cdot (7 \cdot 8) \cdot 10 &= 1 \cdot 1 \cdot 1 \cdot 1 \cdot 1 \cdot -1 \\ &= -1 \pmod{11}. \end{aligned}$$

This reasoning generalizes to any prime mod. Therefore $(p - 1)! = -1 \pmod{p}$.

Supplementary Problem 4.11

Suppose that $/(2a)(\text{mod } n)$ exists. (For purposes of this argument, assume $n > 4$.) All statements below are mod n statements.

x_0 is a solution of the equation $ax^2 + bx + c = 0$

$$\Leftrightarrow \quad ax_0^{\ 2} + bx_0 + c = 0$$

$$\Leftrightarrow \quad ax_0^{\ 2} + bx_0 = -c$$

$$\Leftrightarrow \quad 4a^2x_0^{\ 2} + 4abx_0 = -4ac$$

$$\Leftrightarrow \quad 4a^2x_0^{\ 2} + 4abx_0 + b^2 = b^2 - 4ac$$

$$\Leftrightarrow \quad (2ax_0 + b)^2 = b^2 - 4ac,$$

which is true if and only if $2ax_0 + b$ is a square root of $b^2 - 4ac$

$\Leftrightarrow$ $2ax_0 = -b + k$ where k is a square root of $b^2 - 4ac$

$\Leftrightarrow$ $x_0 = (-b + k)/(2a)$.

Note that if $/(2a)$ exists then $/2$ exists (since if $(2a)b = 1$, then $2(ab) = 1$) and hence $/4 = /2 \cdot /2$ exists.

Supplementary Problem 4.12

1. x^2, $x^2 + x$, $x^2 + 2x$, $x^2 + 1$, $x^2 + x + 1$, $x^2 + 2x + 1$, $x^2 + 2$, $x^2 + x + 2$, and $x^2 + 2x + 2$.

2.

Quadratic	Factorization	Root(s)
x^2	$(x-0)(x-0)$	0
$x^2 + x$	$x(x+1)$ or $x(x-2)$	0, 2
$x^2 + 2x$	$x(x+2)$ or $x(x-1)$	0, 1
$x^2 + x + 1$	$(x+2)(x+2)$ or $(x-1)^2$	1
$x^2 + 2x + 1$	$(x+1)(x+1)$ or $(x-2)^2$	2
$x^2 + 2$	$(x+1)(x+2)$ or $(x-2)(x-1)$	2, 1

4.

a	$-a$	$/a$
0	0	—
1	2	1
2	1	2
i	$2i$	$2i$
$1+i$	$2+2i$	$2+i$
$2+i$	$1+2i$	$1+i$
$2i$	i	i
$1+2i$	$2+i$	$2+2i$
$2+2i$	$1+i$	$1+2i$

5. $x^2 + 1 = (x + i)(x - i) = (x - 2i)(x - i)$. Roots: $2i$ and i
$x^2 + x + 2 = (x - [1 + i])(x - [1 + 2i])$. Roots: $1 + i$ and $1 + 2i$
$x^2 + 2x + 2 = (x - [2 + i])(x - [2 + 2i])$. Roots: $2 + i$ and $2 + 2i$

Supplementary Problem 5.1

Base 12: Suppose that k is a whole number. Then, $r_n(k)$ is the mod n remainder of the last t digits of k (expressed in base 12) if and only if $n \mid 12^t$, that is, if and only if n is of the form $2^m \cdot 3^n$ where $m \leq 2t$ and $n \leq t$ (since $12 = 2^2 \cdot 3$). We need consider precisely the last t digits (and no fewer) if $m = 2t$ and $n \leq t$ or $n = t$ and $m \leq 2t$.

Base 30: Suppose that k is a whole number. Then, $r_n(k)$ is the mod n remainder of the last t digits of k (expressed in base 30) if and only if $n \mid 30^t$ if and only if n is of the form $2^m \cdot 3^n \cdot 5^q$ where m, n, and q are less than or equal to t. We need consider precisely the last t digits (and no fewer) if at least one of m, n, or q equals t.

Supplementary Problem 5.2

Base 9: Suppose that $n > 1$. For any whole number k, $r_n(k)$ is equal to the mod n sum of the digits of k, where k is expressed as a base 9 numeral, if and only if $n \mid 8$. In this case, n can be 2, 4, or 8. In general, for any whole number k, $r_n(k)$ is equal to the mod n sum of the digits of k, where k is expressed as a base b numeral if and only if $n \mid (b - 1)$.

Supplementary Problem 5.3

The solution is given in the text.

Supplementary Problem 5.4

$$
\begin{aligned}
r(35761^{2542}) &= 5^4 && \pmod 7 \\
&= 2. && \pmod 7 \\
r(70146^{66245}) &= 6^{\text{odd power}} && \pmod 7 \\
&= 6. && \pmod 7
\end{aligned}
$$

Supplementary Problem 5.5

$$r(31453^{31259}) = 3^3 \pmod 5$$
$$= 2. \pmod 5$$
$$r(6734521^{215424}) = 2^4 \pmod{11}$$
$$= 5. \pmod{11}$$
$$r(254304^{403}) = 4^3 \pmod{25}$$
$$= 14. \pmod{25}$$

(Note: $U_{25,}$ the set of units of $\mathbf{Z}_{25,}$ has 20 elements, and 4 is one of them. Hence $4^{20} = 1 \pmod{25}$.)

Supplementary Problem 5.6

1. $\phi(p) = p - 1;\ \phi(p^2) = p(p-1);$

   $\phi(p^3) = p^2(p-1);\ \phi(p^k) = p^{k-1}(p-1).$
2. $\phi(pq) = (p-1)(q-1);\ \phi(p^2q) = p(p-1)(q-1);$

   $\phi(p^3q^2) = p^2(p-1)q(q-1);$

   $\phi(p^rq^s) = p^{r-1}(p-1)q^{s-1}(q-1).$
3. $\phi(p_1p_2p_3) = (p_1-1)(p_2-1)(p_3-1);$

   $\phi({p_1}^2{p_2}^3p_3) = p_1(p_1-1){p_2}^2(p_2-1)(p_3-1);$

   $\phi({p_1}^{k_1}{p_2}^{k_2}{p_3}^{k_3}\ldots{p_j}^{k_j}) =$

   $${p_1}^{k_1-1}(p_1-1){p_2}^{k_2-1}(p_2-1){p_3}^{k_3-1}(p_3-1)\ldots{p_j}^{k_j-1}(p_j-1).$$
4. $45 = 3^2(5)$. So, $\phi(45) = 3(3-1)(4) = 24$. 7 is relatively prime to 45, so $7^{24} = 1 \pmod{45}$.

   $1800 = (2^3)(3^2)(5^2)$. So $\phi(1800) = 2^2(2-1)3(3-1)5(5-1) = 480$. 31 is relatively prime to 1800, so $31^{480} = 1 \pmod{1800}$.

Supplementary Problem 6.1

1. $a(x_0 + bt) + b(y_0 - at) = ax_0 + by_0 = k.$
2. Suppose that $ax_0 + by_0 = k$, $ax_1 + by_1 = k$, and $\gcd(a, b) = 1$. Then, subtracting the first two equations, we obtain:

$$ax_0 - ax_1 + by_0 - by_1 = 0$$
$$a(x_0 - x_1) + b(y_0 - y_1) = 0.$$

Now, $a \mid a(x_0 - x_1)$, so $a \mid b(y_0 - y_1)$. But a and b are relatively prime. Hence $a \mid (y_0 - y_1)$. Thus, there exists an integer t such that

$at = y_0 - y_1$. So, $y_1 = y_0 - at$. Substituting this into the second equation above, we obtain:

$$a(x_0 - x_1) + b(y_0 - (y_0 - at)) = 0$$

$$a(x_0 - x_1) + b(at) = 0.$$

Cancelling a's we get: $$x_0 - x_1 + bt = 0$$

$$x_1 = x_0 + bt.$$

Supplementary Problem 6.2

1. There are 6 impossible scores, 1, 2, 4, 5, 8, and 11;
2. There are $(a-1)(b-1)/2$ impossible scores, k, and the largest one is $(a-1)(b-1)-1$. Note that if a and b are not relatively prime, there are infinitely many impossible scores.
3. $(a-1)(b-1) = 70$. Thus the largest impossible score is $70 - 1 = 69$. There are 4 ways to factor 70, but only one for which a and b are relatively prime and both larger than 2. Therefore $a = 8$ and $b = 11$ (or vice versa.)

Supplementary Problem 7.1

When y is nonnegative, it is the number of times you must circle the board (that is, hit or pass [0] again) to reach the desired square in x moves. When $x = 3, y = 1$. If you move <9> 3 times on a 21 board, you will have circled the board once. (You hit [9], [18], then pass [0] to reach [6].) However, when $x = -4$ then $y = -2$ has a slightly different meaning. It means that if you move <9> 4 times to the *left*, you will circle the board twice before landing on [6], *where the first time you move you are considered to be 'passing' [0].* (You pass [0], hit [12], [3], then pass [0] a second time to hit [15], and then reach [6].)

Supplementary Problem 7.2

Suppose that $\gcd(a, n) \mid b$, where a and b are elements of $\mathbf{Z}_n$. Then there are g = $\gcd(a, n)$ elements of $\mathbf{Z}_n$ which are solutions to $ax = b \pmod{n}$. If x_0 is one solution and g > 1, then the others are $x_0 + n/g$, $x_0 + 2n/g$, . . . , $x_0 + (g-1)n/g \pmod{n}$.

Supplementary Problem 7.3

If x is the minimum number of moves, then $2r(x) = 1 \pmod 5$, $5r(x) = 2 \pmod{11}$ and $r(x) = 15 \pmod{18}$. So, $r(x) = 3 \pmod 5$, $r(x) = 7 \pmod{11}$ and $r(x) = 15 \pmod{18}$.

Combining the first two equations we obtain: $r(x) = 3 \pmod 5$, so x is of the form $3 + 5k$. But, $r(x) = 7 \pmod{11}$, so $r(3 + 5k) = 7 \pmod{11}$. Hence $r(k) = 3 \pmod{11}$. Thus k is of the form $3 + 11j$ and therefore, x is of the form $3 + 5k = 3 + 5(3 + 11j) = 18 + 55j$.

Combining this with the third equation, we obtain the equation: $r(18 + 55j) = 15 \pmod{18}$. Hence, $r(j) = 15 \pmod{18}$. The minimum j is 15. Letting $j = 15$, it follows that $x = 18 + 55(15) = 843$.

Challenger

In order for the equations $r(x)r = k \pmod m$ and $r(x)u = j \pmod n$ to have a simultaneous solution, each must individually have roots. That is, we need $\gcd(r, m) \mid k$ and $\gcd(u, n) \mid j$. So, assume that the first equation reduces to $r(x) = a \pmod m$ and the second equation reduces to $r(x) = b \pmod n$. Claim: These equations have a simultaneous solution if and only if $r(a) = r(b) \bmod \gcd(m, n)$. In particular, if m and n are relatively prime, there are simultaneous solutions.

For example, there is no simultaneous solution to the equations $x = 8 \pmod{15}$ and $x = 12 \pmod{18}$ since $r(8)$ doesn't equal $r(12)$ mod 3. (3 is the gcd of 15 and 18). But there are simultaneous solutions to the equations $r(x) = 14 \pmod{18}$ and $r(x) = 20 \pmod{30}$ since $r(14) = r(20) \pmod 6$.

Proof: Let $a \in \mathbf{Z}_m$, $b \in \mathbf{Z}_n$, and $g = \gcd(m, n)$. Assume that there exists x such that $r(x)=a \pmod m$ and $r(x) = b \pmod n$. Then $r(x - a) = 0 \pmod m$ and $r(x - b) = 0 \pmod n$.

So there exist t and s such that $x - a = tm$ and $x - b = sn$.

So
$$a + tm = b + sn$$
$$tm - sn = b - a.$$

Thus $g \mid (b - a)$. Therefore, $r(b-a) = 0 \pmod g$. Thus, $r(b) = r(a) \pmod g$.

Conversely, suppose that $r(b) = r(a) \pmod g$. Then $g \mid (b - a)$. So there exist t and s such that $tm - sn = b - a$. Let $x = tm + a = sn + b$. Then $r(x) = a \pmod m$ and $r(x) = b \pmod n$.

Commentary

by Richard Singer, Webster University

In a 19th century book on Quaternions, Hamilton gives a justification for the study of a non-commutative algebraic structure. Contemporary mathematicians would feel no need for justification. We would simply study any algebraic structure that interested us, and to interest others we would appeal to elegance or mathematical significance. The 19th century way of thinking seems so alien that we would hardly be aware of it, except for the fact that we encounter something like it in the majority of our students and perhaps somewhat in our own attitudes toward logic and set theory.

It is difficult to articulate the difference between the typical 19th and 20th century attitudes towards mathematics. For now we can summarize this difference by saying that they focused on subject matter, while we focus on a special way of studying abstract conceptual nets. They had concerns about the appropriateness of what we would have considered legitimate subject matter for mathematics. We regard anything which we can imagine formalizing as appropriate for mathematical study. In discussing the difference between 19th and 20th century mathematics with several mathematicians, they all agreed in principle with a *conceptual* versus *paraconceptual* distinction.

Conceptual versus paraconceptual A *conceptual net* is a collection of related concepts and conceptual distinctions that can be used to think and communicate about some realm of interest. Pure conceptual study involves the logical study of some realm which is itself a conceptual net. To be engaged in conceptual study is to focus on presenting, clarifying, and refining concepts. This always focuses on relationships among concepts within some conceptual net, either to clarify or refine the net or to create a new or alternative one. The goal is to obtain better knowledge and more skill in regards to the net being considered, and the only claims made are about the net under consideration.

On the other hand, most types of studies make what could be called *paraconceptual* claims, that is, claims that use concepts to talk about some states of affairs that are not being considered part of the conceptual net. Consider the game Gin Rummy. The jack of clubs is a club with the rank of Jack and with a value of 10. To study this game it is clear that a person must be able to use concepts like suit, rank, value, color of a suit, and same rank. Conceptually, a set is at least three cards of the same rank. Fairly trivial reasoning reveals the conceptual claim that any set must contain at least one red card. Mere conceptual analysis of the game rules will not demonstrate that it is poor strategy to keep large cards too long. This is a paraconceptual claim, rather than a conceptual one relative to this net. In a conceptual net that included probability as well as the game rules, some claims about strategy could be construed as conceptual rather than as paraconceptual.

Expectations for students There is a radical difference in our expectations for those we consider potential mathematicians and a more general audience who we think will merely use mathematics. The latter are seldom exposed to mathematics as a conceptual net. For them mathematics is generally taught as a collection of algorithms, with applications to word problems that students generally dislike and find rather pointless. They are told that mathematics has important applications or that they will need this material in more advanced courses. Since most of the mathematics taught in secondary school and early in college have applications that most students lack the background to appreciate, this hardly provides perspective either on the internal nature of mathematics or on its place in the broader scheme of our intellectual tradition.

Students majoring in mathematics are exposed to many abstract definitions and proofs, and while the emphasis is placed on knowing results, students are also expected to absorb our contemporary way of thinking. Unfortunately neither this nor the difference between the contemporary and older perspective on mathematics is examined, except on rare occasions. As a result only a minority of such students gain much perspective on the nature of contemporary mathematics, and few have any explicit historical perspective on mathematics and its epistemological significance.

The Theory of Remainders The mathematics in the *Theory of Remainders* focuses on the study of modular structures in a way that challenges the students to adopt a 20th century perspective and to do some significant thinking from this perspective. However, it accomplishes this without requiring the student to have an extensive mathematical background. Modular structures are presented as algebraic structures in their own right rather than from the older perspective of congruences.

- Since these structures are rings, they have the properties needed to use and reinforce many standard ideas of ordinary algebra. The fact that these ideas are inherent in the conceptual net, rather than somehow being specialized to subsystems of the complex numbers, is well illustrated.

- Since some modular structures are not cancellation rings, the presence of zero divisors can be contrasted with their absence in more familiar systems, and this can be related to finding solutions of equations in both types of structures. The fact that these structures are not ordered rings gives a plausible reason to examine the concept of negative numbers from the broader perspective of additive inverses. That some of these structures are fields illustrates the algebraic nature of multiplicative inverses.

In addition to comparing and contrasting modular to more familiar structures, this book uses the remainder function to solve problems about integers by mapping them to simpler problems in modular structures. This illustrates the concept and use of a morphism at a very elementary level.

The above comments indicate only a few of the specific ways that this book focuses on a contemporary approach to mathematics in an elementary way and in a manner that is appropriate for an introduction to mathematics as the study of conceptual nets. In particular it deals with a unified net of concepts in an elementary but challenging manner. Supporting computations take imagination but are not tedious. They compare and contrast easily with ideas from ordinary algebra.

There are a number of materials which could be used to introduce students to mathematics as a conceptual net rather than as a study of paraconceptual claims. These include Boolean Algebra, transfinite ordinals, Gaussian integers, quaternions, functions algebras, symmetry groups, matrix algebra, nonstandard real numbers, and elementary mathematical logic. Some materials are being developed in most of these areas. None of these areas contain all the advantages of the modular structures, although the study of Boolean Algebra and selected subjects from some of the other areas could make an excellent follow-up course. Furthermore, a study of a number of books and articles on the history and nature of mathematics also comes to mind. For example, the Newman volumes on the world of mathematics, *Mathematics in Western Culture* by Kline, *What is Mathematics* by Courant and Robbins, and the Halmos article on mathematics as a creative art all contain some excellent material which could be used to supplement either a beginning or follow-up study of mathematics as the rigorous exploration of conceptual nets.

Expanding perspectives Many arguments about ordinary matters are sometimes said to hinge on semantic differences. In most cases I would say they are bogged down in vague or primitive conceptual nets and in the inability of people to clarify nets they are using. This is often complicated by an emotional interest in the outcome of the argument, so that winning a point becomes more important than finding the core of the disagreement. Where better to begin to become sensitive to the role of our concepts than in mathematics, where this awareness first evolved and where the nets can be most easily articulated? Unfortunately this is not the way most people encounter mathematics. *The Theory of Remainders* can provide a significant step in seeing mathematics as a conceptual net and in understanding the difference between conceptual and paraconceptual claims—an excellent beginning in expanding people's perspective on mathematics.

Index

A CATALOG OF SELECTED

DOVER BOOKS

IN SCIENCE AND MATHEMATICS

Astronomy

BURNHAM'S CELESTIAL HANDBOOK, Robert Burnham, Jr. Thorough guide to the stars beyond our solar system. Exhaustive treatment. Alphabetical by constellation: Andromeda to Cetus in Vol. 1; Chamaeleon to Orion in Vol. 2; and Pavo to Vulpecula in Vol. 3. Hundreds of illustrations. Index in Vol. 3. 2,000pp. 6⅛ x 9¼.
Vol. I: 23567-X
Vol. II: 23568-8
Vol. III: 23673-0

EXPLORING THE MOON THROUGH BINOCULARS AND SMALL TELESCOPES, Ernest H. Cherrington, Jr. Informative, profusely illustrated guide to locating and identifying craters, rills, seas, mountains, other lunar features. Newly revised and updated with special section of new photos. Over 100 photos and diagrams. 240pp. 8¼ x 11. 24491-1

THE EXTRATERRESTRIAL LIFE DEBATE, 1750–1900, Michael J. Crowe. First detailed, scholarly study in English of the many ideas that developed from 1750 to 1900 regarding the existence of intelligent extraterrestrial life. Examines ideas of Kant, Herschel, Voltaire, Percival Lowell, many other scientists and thinkers. 16 illustrations. 704pp. 5⅜ x 8½. 40675-X

THEORIES OF THE WORLD FROM ANTIQUITY TO THE COPERNICAN REVOLUTION, Michael J. Crowe. Newly revised edition of an accessible, enlightening book recreates the change from an earth-centered to a sun-centered conception of the solar system. 242pp. 5⅜ x 8½. 41444-2

A HISTORY OF ASTRONOMY, A. Pannekoek. Well-balanced, carefully reasoned study covers such topics as Ptolemaic theory, work of Copernicus, Kepler, Newton, Eddington's work on stars, much more. Illustrated. References. 521pp. 5⅜ x 8½. 65994-1

A COMPLETE MANUAL OF AMATEUR ASTRONOMY: Tools and Techniques for Astronomical Observations, P. Clay Sherrod with Thomas L. Koed. Concise, highly readable book discusses: selecting, setting up and maintaining a telescope; amateur studies of the sun; lunar topography and occultations; observations of Mars, Jupiter, Saturn, the minor planets and the stars; an introduction to photoelectric photometry; more. 1981 ed. 124 figures. 26 halftones. 37 tables. 335pp. 6½ x 9¼. 42820-6

AMATEUR ASTRONOMER'S HANDBOOK, J. B. Sidgwick. Timeless, comprehensive coverage of telescopes, mirrors, lenses, mountings, telescope drives, micrometers, spectroscopes, more. 189 illustrations. 576pp. 5⅝ x 8¼. (Available in U.S. only.) 24034-7

STARS AND RELATIVITY, Ya. B. Zel'dovich and I. D. Novikov. Vol. 1 of *Relativistic Astrophysics* by famed Russian scientists. General relativity, properties of matter under astrophysical conditions, stars, and stellar systems. Deep physical insights, clear presentation. 1971 edition. References. 544pp. 5⅝ x 8¼. 69424-0

Chemistry

THE SCEPTICAL CHYMIST: The Classic 1661 Text, Robert Boyle. Boyle defines the term "element," asserting that all natural phenomena can be explained by the motion and organization of primary particles. 1911 ed. viii+232pp. 5⅜ x 8½. 42825-7

RADIOACTIVE SUBSTANCES, Marie Curie. Here is the celebrated scientist's doctoral thesis, the prelude to her receipt of the 1903 Nobel Prize. Curie discusses establishing atomic character of radioactivity found in compounds of uranium and thorium; extraction from pitchblende of polonium and radium; isolation of pure radium chloride; determination of atomic weight of radium; plus electric, photographic, luminous, heat, color effects of radioactivity. ii+94pp. 5⅜ x 8½. 42550-9

CHEMICAL MAGIC, Leonard A. Ford. Second Edition, Revised by E. Winston Grundmeier. Over 100 unusual stunts demonstrating cold fire, dust explosions, much more. Text explains scientific principles and stresses safety precautions. 128pp. 5⅜ x 8½. 67628-5

THE DEVELOPMENT OF MODERN CHEMISTRY, Aaron J. Ihde. Authoritative history of chemistry from ancient Greek theory to 20th-century innovation. Covers major chemists and their discoveries. 209 illustrations. 14 tables. Bibliographies. Indices. Appendices. 851pp. 5⅜ x 8½. 64235-6

CATALYSIS IN CHEMISTRY AND ENZYMOLOGY, William P. Jencks. Exceptionally clear coverage of mechanisms for catalysis, forces in aqueous solution, carbonyl- and acyl-group reactions, practical kinetics, more. 864pp. 5⅜ x 8½. 65460-5

ELEMENTS OF CHEMISTRY, Antoine Lavoisier. Monumental classic by founder of modern chemistry in remarkable reprint of rare 1790 Kerr translation. A must for every student of chemistry or the history of science. 539pp. 5⅜ x 8½. 64624-6

THE HISTORICAL BACKGROUND OF CHEMISTRY, Henry M. Leicester. Evolution of ideas, not individual biography. Concentrates on formulation of a coherent set of chemical laws. 260pp. 5⅜ x 8½. 61053-5

A SHORT HISTORY OF CHEMISTRY, J. R. Partington. Classic exposition explores origins of chemistry, alchemy, early medical chemistry, nature of atmosphere, theory of valency, laws and structure of atomic theory, much more. 428pp. 5⅜ x 8½. (Available in U.S. only.) 65977-1

GENERAL CHEMISTRY, Linus Pauling. Revised 3rd edition of classic first-year text by Nobel laureate. Atomic and molecular structure, quantum mechanics, statistical mechanics, thermodynamics correlated with descriptive chemistry. Problems. 992pp. 5⅜ x 8½. 65622-5

FROM ALCHEMY TO CHEMISTRY, John Read. Broad, humanistic treatment focuses on great figures of chemistry and ideas that revolutionized the science. 50 illustrations. 240pp. 5⅜ x 8½. 28690-8

Engineering

DE RE METALLICA, Georgius Agricola. The famous Hoover translation of greatest treatise on technological chemistry, engineering, geology, mining of early modern times (1556). All 289 original woodcuts. 638pp. 6¾ x 11. 60006-8

FUNDAMENTALS OF ASTRODYNAMICS, Roger Bate et al. Modern approach developed by U.S. Air Force Academy. Designed as a first course. Problems, exercises. Numerous illustrations. 455pp. 5⅜ x 8½. 60061-0

DYNAMICS OF FLUIDS IN POROUS MEDIA, Jacob Bear. For advanced students of ground water hydrology, soil mechanics and physics, drainage and irrigation engineering, and more. 335 illustrations. Exercises, with answers. 784pp. 6⅛ x 9¼. 65675-6

THEORY OF VISCOELASTICITY (Second Edition), Richard M. Christensen. Complete, consistent description of the linear theory of the viscoelastic behavior of materials. Problem-solving techniques discussed. 1982 edition. 29 figures. xiv+364pp. 6⅛ x 9¼. 42880-X

MECHANICS, J. P. Den Hartog. A classic introductory text or refresher. Hundreds of applications and design problems illuminate fundamentals of trusses, loaded beams and cables, etc. 334 answered problems. 462pp. 5⅜ x 8½. 60754-2

MECHANICAL VIBRATIONS, J. P. Den Hartog. Classic textbook offers lucid explanations and illustrative models, applying theories of vibrations to a variety of practical industrial engineering problems. Numerous figures. 233 problems, solutions. Appendix. Index. Preface. 436pp. 5⅜ x 8½. 64785-4

STRENGTH OF MATERIALS, J. P. Den Hartog. Full, clear treatment of basic material (tension, torsion, bending, etc.) plus advanced material on engineering methods, applications. 350 answered problems. 323pp. 5⅜ x 8½. 60755-0

A HISTORY OF MECHANICS, René Dugas. Monumental study of mechanical principles from antiquity to quantum mechanics. Contributions of ancient Greeks, Galileo, Leonardo, Kepler, Lagrange, many others. 671pp. 5⅜ x 8½. 65632-2

STABILITY THEORY AND ITS APPLICATIONS TO STRUCTURAL MECHANICS, Clive L. Dym. Self-contained text focuses on Koiter postbuckling analyses, with mathematical notions of stability of motion. Basing minimum energy principles for static stability upon dynamic concepts of stability of motion, it develops asymptotic buckling and postbuckling analyses from potential energy considerations, with applications to columns, plates, and arches. 1974 ed. 208pp. 5⅜ x 8½. 42541-X

METAL FATIGUE, N. E. Frost, K. J. Marsh, and L. P. Pook. Definitive, clearly written, and well-illustrated volume addresses all aspects of the subject, from the historical development of understanding metal fatigue to vital concepts of the cyclic stress that causes a crack to grow. Includes 7 appendixes. 544pp. 5⅜ x 8½. 40927-9

ROCKETS, Robert Goddard. Two of the most significant publications in the history of rocketry and jet propulsion: "A Method of Reaching Extreme Altitudes" (1919) and "Liquid Propellant Rocket Development" (1936). 128pp. 5⅜ x 8½. 42537-1

STATISTICAL MECHANICS: Principles and Applications, Terrell L. Hill. Standard text covers fundamentals of statistical mechanics, applications to fluctuation theory, imperfect gases, distribution functions, more. 448pp. 5⅜ x 8½. 65390-0

ENGINEERING AND TECHNOLOGY 1650–1750: Illustrations and Texts from Original Sources, Martin Jensen. Highly readable text with more than 200 contemporary drawings and detailed engravings of engineering projects dealing with surveying, leveling, materials, hand tools, lifting equipment, transport and erection, piling, bailing, water supply, hydraulic engineering, and more. Among the specific projects outlined–transporting a 50-ton stone to the Louvre, erecting an obelisk, building timber locks, and dredging canals. 207pp. 8⅜ x 11¼. 42232-1

THE VARIATIONAL PRINCIPLES OF MECHANICS, Cornelius Lanczos. Graduate level coverage of calculus of variations, equations of motion, relativistic mechanics, more. First inexpensive paperbound edition of classic treatise. Index. Bibliography. 418pp. 5⅜ x 8½. 65067-7

PROTECTION OF ELECTRONIC CIRCUITS FROM OVERVOLTAGES, Ronald B. Standler. Five-part treatment presents practical rules and strategies for circuits designed to protect electronic systems from damage by transient overvoltages. 1989 ed. xxiv+434pp. 6⅛ x 9¼. 42552-5

ROTARY WING AERODYNAMICS, W. Z. Stepniewski. Clear, concise text covers aerodynamic phenomena of the rotor and offers guidelines for helicopter performance evaluation. Originally prepared for NASA. 537 figures. 640pp. 6⅛ x 9¼. 64647-5

INTRODUCTION TO SPACE DYNAMICS, William Tyrrell Thomson. Comprehensive, classic introduction to space-flight engineering for advanced undergraduate and graduate students. Includes vector algebra, kinematics, transformation of coordinates. Bibliography. Index. 352pp. 5⅜ x 8½. 65113-4

HISTORY OF STRENGTH OF MATERIALS, Stephen P. Timoshenko. Excellent historical survey of the strength of materials with many references to the theories of elasticity and structure. 245 figures. 452pp. 5⅜ x 8½. 61187-6

ANALYTICAL FRACTURE MECHANICS, David J. Unger. Self-contained text supplements standard fracture mechanics texts by focusing on analytical methods for determining crack-tip stress and strain fields. 336pp. 6⅛ x 9¼. 41737-9

STATISTICAL MECHANICS OF ELASTICITY, J. H. Weiner. Advanced, self-contained treatment illustrates general principles and elastic behavior of solids. Part 1, based on classical mechanics, studies thermoelastic behavior of crystalline and polymeric solids. Part 2, based on quantum mechanics, focuses on interatomic force laws, behavior of solids, and thermally activated processes. For students of physics and chemistry and for polymer physicists. 1983 ed. 96 figures. 496pp. 5⅜ x 8½. 42260-7

Mathematics

FUNCTIONAL ANALYSIS (Second Corrected Edition), George Bachman and Lawrence Narici. Excellent treatment of subject geared toward students with background in linear algebra, advanced calculus, physics, and engineering. Text covers introduction to inner-product spaces, normed, metric spaces, and topological spaces; complete orthonormal sets, the Hahn-Banach Theorem and its consequences, and many other related subjects. 1966 ed. 544pp. 6⅛ x 9¼. 40251-7

ASYMPTOTIC EXPANSIONS OF INTEGRALS, Norman Bleistein & Richard A. Handelsman. Best introduction to important field with applications in a variety of scientific disciplines. New preface. Problems. Diagrams. Tables. Bibliography. Index. 448pp. 5⅜ x 8½. 65082-0

VECTOR AND TENSOR ANALYSIS WITH APPLICATIONS, A. I. Borisenko and I. E. Tarapov. Concise introduction. Worked-out problems, solutions, exercises. 257pp. 5⅝ x 8¼. 63833-2

THE ABSOLUTE DIFFERENTIAL CALCULUS (CALCULUS OF TENSORS), Tullio Levi-Civita. Great 20th-century mathematician's classic work on material necessary for mathematical grasp of theory of relativity. 452pp. 5⅝ x 8¼. 63401-9

AN INTRODUCTION TO ORDINARY DIFFERENTIAL EQUATIONS, Earl A. Coddington. A thorough and systematic first course in elementary differential equations for undergraduates in mathematics and science, with many exercises and problems (with answers). Index. 304pp. 5⅜ x 8½. 65942-9

FOURIER SERIES AND ORTHOGONAL FUNCTIONS, Harry F. Davis. An incisive text combining theory and practical example to introduce Fourier series, orthogonal functions and applications of the Fourier method to boundary-value problems. 570 exercises. Answers and notes. 416pp. 5⅜ x 8½. 65973-9

COMPUTABILITY AND UNSOLVABILITY, Martin Davis. Classic graduate-level introduction to theory of computability, usually referred to as theory of recurrent functions. New preface and appendix. 288pp. 5⅜ x 8½. 61471-9

ASYMPTOTIC METHODS IN ANALYSIS, N. G. de Bruijn. An inexpensive, comprehensive guide to asymptotic methods–the pioneering work that teaches by explaining worked examples in detail. Index. 224pp. 5⅜ x 8½ 64221-6

APPLIED COMPLEX VARIABLES, John W. Dettman. Step-by-step coverage of fundamentals of analytic function theory–plus lucid exposition of five important applications: Potential Theory; Ordinary Differential Equations; Fourier Transforms; Laplace Transforms; Asymptotic Expansions. 66 figures. Exercises at chapter ends. 512pp. 5⅜ x 8½. 64670-X

INTRODUCTION TO LINEAR ALGEBRA AND DIFFERENTIAL EQUATIONS, John W. Dettman. Excellent text covers complex numbers, determinants, orthonormal bases, Laplace transforms, much more. Exercises with solutions. Undergraduate level. 416pp. 5⅜ x 8½. 65191-6

CALCULUS OF VARIATIONS WITH APPLICATIONS, George M. Ewing. Applications-oriented introduction to variational theory develops insight and promotes understanding of specialized books, research papers. Suitable for advanced undergraduate/graduate students as primary, supplementary text. 352pp. 5⅜ x 8½. 64856-7

COMPLEX VARIABLES, Francis J. Flanigan. Unusual approach, delaying complex algebra till harmonic functions have been analyzed from real variable viewpoint. Includes problems with answers. 364pp. 5⅜ x 8½. 61388-7

AN INTRODUCTION TO THE CALCULUS OF VARIATIONS, Charles Fox. Graduate-level text covers variations of an integral, isoperimetrical problems, least action, special relativity, approximations, more. References. 279pp. 5⅜ x 8½. 65499-0

COUNTEREXAMPLES IN ANALYSIS, Bernard R. Gelbaum and John M. H. Olmsted. These counterexamples deal mostly with the part of analysis known as "real variables." The first half covers the real number system, and the second half encompasses higher dimensions. 1962 edition. xxiv+198pp. 5⅜ x 8½. 42875-3

CATASTROPHE THEORY FOR SCIENTISTS AND ENGINEERS, Robert Gilmore. Advanced-level treatment describes mathematics of theory grounded in the work of Poincaré, R. Thom, other mathematicians. Also important applications to problems in mathematics, physics, chemistry, and engineering. 1981 edition. References. 28 tables. 397 black-and-white illustrations. xvii+666pp. 6⅛ x 9¼. 67539-4

INTRODUCTION TO DIFFERENCE EQUATIONS, Samuel Goldberg. Exceptionally clear exposition of important discipline with applications to sociology, psychology, economics. Many illustrative examples; over 250 problems. 260pp. 5⅜ x 8½. 65084-7

NUMERICAL METHODS FOR SCIENTISTS AND ENGINEERS, Richard Hamming. Classic text stresses frequency approach in coverage of algorithms, polynomial approximation, Fourier approximation, exponential approximation, other topics. Revised and enlarged 2nd edition. 721pp. 5⅜ x 8½. 65241-6

INTRODUCTION TO NUMERICAL ANALYSIS (2nd Edition), F. B. Hildebrand. Classic, fundamental treatment covers computation, approximation, interpolation, numerical differentiation and integration, other topics. 150 new problems. 669pp. 5⅜ x 8½. 65363-3

THREE PEARLS OF NUMBER THEORY, A. Y. Khinchin. Three compelling puzzles require proof of a basic law governing the world of numbers. Challenges concern van der Waerden's theorem, the Landau-Schnirelmann hypothesis and Mann's theorem, and a solution to Waring's problem. Solutions included. 64pp. 5⅜ x 8½. 40026-3

THE PHILOSOPHY OF MATHEMATICS: An Introductory Essay, Stephan Körner. Surveys the views of Plato, Aristotle, Leibniz & Kant concerning propositions and theories of applied and pure mathematics. Introduction. Two appendices. Index. 198pp. 5⅜ x 8½. 25048-2

INTRODUCTORY REAL ANALYSIS, A.N. Kolmogorov, S. V. Fomin. Translated by Richard A. Silverman. Self-contained, evenly paced introduction to real and functional analysis. Some 350 problems. 403pp. 5⅜ x 8½. 61226-0

APPLIED ANALYSIS, Cornelius Lanczos. Classic work on analysis and design of finite processes for approximating solution of analytical problems. Algebraic equations, matrices, harmonic analysis, quadrature methods, more. 559pp. 5⅜ x 8½. 65656-X

AN INTRODUCTION TO ALGEBRAIC STRUCTURES, Joseph Landin. Superb self-contained text covers "abstract algebra": sets and numbers, theory of groups, theory of rings, much more. Numerous well-chosen examples, exercises. 247pp. 5⅜ x 8½. 65940-2

QUALITATIVE THEORY OF DIFFERENTIAL EQUATIONS, V. V. Nemytskii and V.V. Stepanov. Classic graduate-level text by two prominent Soviet mathematicians covers classical differential equations as well as topological dynamics and ergodic theory. Bibliographies. 523pp. 5⅜ x 8½. 65954-2

THEORY OF MATRICES, Sam Perlis. Outstanding text covering rank, nonsingularity and inverses in connection with the development of canonical matrices under the relation of equivalence, and without the intervention of determinants. Includes exercises. 237pp. 5⅜ x 8½. 66810-X

INTRODUCTION TO ANALYSIS, Maxwell Rosenlicht. Unusually clear, accessible coverage of set theory, real number system, metric spaces, continuous functions, Riemann integration, multiple integrals, more. Wide range of problems. Undergraduate level. Bibliography. 254pp. 5⅜ x 8½. 65038-3

MODERN NONLINEAR EQUATIONS, Thomas L. Saaty. Emphasizes practical solution of problems; covers seven types of equations. ". . . a welcome contribution to the existing literature. . . . "–*Math Reviews*. 490pp. 5⅜ x 8½. 64232-1

MATRICES AND LINEAR ALGEBRA, Hans Schneider and George Phillip Barker. Basic textbook covers theory of matrices and its applications to systems of linear equations and related topics such as determinants, eigenvalues, and differential equations. Numerous exercises. 432pp. 5⅜ x 8½. 66014-1

MATHEMATICS APPLIED TO CONTINUUM MECHANICS, Lee A. Segel. Analyzes models of fluid flow and solid deformation. For upper-level math, science, and engineering students. 608pp. 5⅜ x 8½. 65369-2

ELEMENTS OF REAL ANALYSIS, David A. Sprecher. Classic text covers fundamental concepts, real number system, point sets, functions of a real variable, Fourier series, much more. Over 500 exercises. 352pp. 5⅜ x 8½. 65385-4

SET THEORY AND LOGIC, Robert R. Stoll. Lucid introduction to unified theory of mathematical concepts. Set theory and logic seen as tools for conceptual understanding of real number system. 496pp. 5⅜ x 8¼. 63829-4

TENSOR CALCULUS, J.L. Synge and A. Schild. Widely used introductory text covers spaces and tensors, basic operations in Riemannian space, non-Riemannian spaces, etc. 324pp. 5⅜ x 8¼. 63612-7

ORDINARY DIFFERENTIAL EQUATIONS, Morris Tenenbaum and Harry Pollard. Exhaustive survey of ordinary differential equations for undergraduates in mathematics, engineering, science. Thorough analysis of theorems. Diagrams. Bibliography. Index. 818pp. 5⅜ x 8½. 64940-7

INTEGRAL EQUATIONS, F. G. Tricomi. Authoritative, well-written treatment of extremely useful mathematical tool with wide applications. Volterra Equations, Fredholm Equations, much more. Advanced undergraduate to graduate level. Exercises. Bibliography. 238pp. 5⅜ x 8½. 64828-1

FOURIER SERIES, Georgi P. Tolstov. Translated by Richard A. Silverman. A valuable addition to the literature on the subject, moving clearly from subject to subject and theorem to theorem. 107 problems, answers. 336pp. 5⅜ x 8½. 63317-9

INTRODUCTION TO MATHEMATICAL THINKING, Friedrich Waismann. Examinations of arithmetic, geometry, and theory of integers; rational and natural numbers; complete induction; limit and point of accumulation; remarkable curves; complex and hypercomplex numbers, more. 1959 ed. 27 figures. xii+260pp. 5⅜ x 8½. 42804-4

POPULAR LECTURES ON MATHEMATICAL LOGIC, Hao Wang. Noted logician's lucid treatment of historical developments, set theory, model theory, recursion theory and constructivism, proof theory, more. 3 appendixes. Bibliography. 1981 ed. ix+283pp. 5⅜ x 8½. 67632-3

CALCULUS OF VARIATIONS, Robert Weinstock. Basic introduction covering isoperimetric problems, theory of elasticity, quantum mechanics, electrostatics, etc. Exercises throughout. 326pp. 5⅜ x 8½. 63069-2

THE CONTINUUM: A Critical Examination of the Foundation of Analysis, Hermann Weyl. Classic of 20th-century foundational research deals with the conceptual problem posed by the continuum. 156pp. 5⅜ x 8½. 67982-9

CHALLENGING MATHEMATICAL PROBLEMS WITH ELEMENTARY SOLUTIONS, A. M. Yaglom and I. M. Yaglom. Over 170 challenging problems on probability theory, combinatorial analysis, points and lines, topology, convex polygons, many other topics. Solutions. Total of 445pp. 5⅜ x 8½. Two-vol. set.
Vol. I: 65536-9 Vol. II: 65537-7

INTRODUCTION TO PARTIAL DIFFERENTIAL EQUATIONS WITH APPLICATIONS, E. C. Zachmanoglou and Dale W. Thoe. Essentials of partial differential equations applied to common problems in engineering and the physical sciences. Problems and answers. 416pp. 5⅜ x 8½. 65251-3

THE THEORY OF GROUPS, Hans J. Zassenhaus. Well-written graduate-level text acquaints reader with group-theoretic methods and demonstrates their usefulness in mathematics. Axioms, the calculus of complexes, homomorphic mapping, *p*-group theory, more. 276pp. 5⅜ x 8½. 40922-8

Math–Decision Theory, Statistics, Probability

ELEMENTARY DECISION THEORY, Herman Chernoff and Lincoln E. Moses. Clear introduction to statistics and statistical theory covers data processing, probability and random variables, testing hypotheses, much more. Exercises. 364pp. 5⅜ x 8½. 65218-1

STATISTICS MANUAL, Edwin L. Crow et al. Comprehensive, practical collection of classical and modern methods prepared by U.S. Naval Ordnance Test Station. Stress on use. Basics of statistics assumed. 288pp. 5⅜ x 8½. 60599-X

SOME THEORY OF SAMPLING, William Edwards Deming. Analysis of the problems, theory, and design of sampling techniques for social scientists, industrial managers, and others who find statistics important at work. 61 tables. 90 figures. xvii +602pp. 5⅜ x 8½. 64684-X

LINEAR PROGRAMMING AND ECONOMIC ANALYSIS, Robert Dorfman, Paul A. Samuelson and Robert M. Solow. First comprehensive treatment of linear programming in standard economic analysis. Game theory, modern welfare economics, Leontief input-output, more. 525pp. 5⅜ x 8½. 65491-5

PROBABILITY: An Introduction, Samuel Goldberg. Excellent basic text covers set theory, probability theory for finite sample spaces, binomial theorem, much more. 360 problems. Bibliographies. 322pp. 5⅜ x 8½. 65252-1

GAMES AND DECISIONS: Introduction and Critical Survey, R. Duncan Luce and Howard Raiffa. Superb nontechnical introduction to game theory, primarily applied to social sciences. Utility theory, zero-sum games, n-person games, decision-making, much more. Bibliography. 509pp. 5⅜ x 8½. 65943-7

INTRODUCTION TO THE THEORY OF GAMES, J. C. C. McKinsey. This comprehensive overview of the mathematical theory of games illustrates applications to situations involving conflicts of interest, including economic, social, political, and military contexts. Appropriate for advanced undergraduate and graduate courses; advanced calculus a prerequisite. 1952 ed. x+372pp. 5⅜ x 8½. 42811-7

FIFTY CHALLENGING PROBLEMS IN PROBABILITY WITH SOLUTIONS, Frederick Mosteller. Remarkable puzzlers, graded in difficulty, illustrate elementary and advanced aspects of probability. Detailed solutions. 88pp. 5⅜ x 8½. 65355-2

PROBABILITY THEORY: A Concise Course, Y. A. Rozanov. Highly readable, self-contained introduction covers combination of events, dependent events, Bernoulli trials, etc. 148pp. 5⅜ x 8¼. 63544-9

STATISTICAL METHOD FROM THE VIEWPOINT OF QUALITY CONTROL, Walter A. Shewhart. Important text explains regulation of variables, uses of statistical control to achieve quality control in industry, agriculture, other areas. 192pp. 5⅜ x 8½. 65232-7

Math–Geometry and Topology

ELEMENTARY CONCEPTS OF TOPOLOGY, Paul Alexandroff. Elegant, intuitive approach to topology from set-theoretic topology to Betti groups; how concepts of topology are useful in math and physics. 25 figures. 57pp. 5⅜ x 8½. 60747-X

COMBINATORIAL TOPOLOGY, P. S. Alexandrov. Clearly written, well-organized, three-part text begins by dealing with certain classic problems without using the formal techniques of homology theory and advances to the central concept, the Betti groups. Numerous detailed examples. 654pp. 5⅜ x 8½. 40179-0

EXPERIMENTS IN TOPOLOGY, Stephen Barr. Classic, lively explanation of one of the byways of mathematics. Klein bottles, Moebius strips, projective planes, map coloring, problem of the Koenigsberg bridges, much more, described with clarity and wit. 43 figures. 210pp. 5⅜ x 8½. 25933-1

CONFORMAL MAPPING ON RIEMANN SURFACES, Harvey Cohn. Lucid, insightful book presents ideal coverage of subject. 334 exercises make book perfect for self-study. 55 figures. 352pp. 5⅝ x 8¼. 64025-6

THE GEOMETRY OF RENÉ DESCARTES, René Descartes. The great work founded analytical geometry. Original French text, Descartes's own diagrams, together with definitive Smith-Latham translation. 244pp. 5⅜ x 8½. 60068-8

PRACTICAL CONIC SECTIONS: The Geometric Properties of Ellipses, Parabolas and Hyperbolas, J. W. Downs. This text shows how to create ellipses, parabolas, and hyperbolas. It also presents historical background on their ancient origins and describes the reflective properties and roles of curves in design applications. 1993 ed. 98 figures. xii+100pp. 6½ x 9¼. 42876-1

THE THIRTEEN BOOKS OF EUCLID'S ELEMENTS, translated with introduction and commentary by Thomas L. Heath. Definitive edition. Textual and linguistic notes, mathematical analysis. 2,500 years of critical commentary. Unabridged. 1,414pp. 5⅜ x 8½.. Three-vol. set. Vol. I: 60088-2 Vol. II: 60089-0 Vol. III: 60090-4

GEOMETRY OF COMPLEX NUMBERS, Hans Schwerdtfeger. Illuminating, widely praised book on analytic geometry of circles, the Moebius transformation, and two-dimensional non-Euclidean geometries. 200pp. 5⅝ x 8¼. 63830-8

DIFFERENTIAL GEOMETRY, Heinrich W. Guggenheimer. Local differential geometry as an application of advanced calculus and linear algebra. Curvature, transformation groups, surfaces, more. Exercises. 62 figures. 378pp. 5⅜ x 8½. 63433-7

CURVATURE AND HOMOLOGY: Enlarged Edition, Samuel I. Goldberg. Revised edition examines topology of differentiable manifolds; curvature, homology of Riemannian manifolds; compact Lie groups; complex manifolds; curvature, homology of Kaehler manifolds. New Preface. Four new appendixes. 416pp. 5⅜ x 8½. 40207-X

History of Math

THE WORKS OF ARCHIMEDES, Archimedes (T. L. Heath, ed.). Topics include the famous problems of the ratio of the areas of a cylinder and an inscribed sphere; the measurement of a circle; the properties of conoids, spheroids, and spirals; and the quadrature of the parabola. Informative introduction. clxxxvi+326pp; supplement, 52pp. 5⅜ x 8½. 42084-1

A SHORT ACCOUNT OF THE HISTORY OF MATHEMATICS, W. W. Rouse Ball. One of clearest, most authoritative surveys from the Egyptians and Phoenicians through 19th-century figures such as Grassman, Galois, Riemann. Fourth edition. 522pp. 5⅜ x 8½. 20630-0

THE HISTORY OF THE CALCULUS AND ITS CONCEPTUAL DEVELOPMENT, Carl B. Boyer. Origins in antiquity, medieval contributions, work of Newton, Leibniz, rigorous formulation. Treatment is verbal. 346pp. 5⅜ x 8½. 60509-4

THE HISTORICAL ROOTS OF ELEMENTARY MATHEMATICS, Lucas N. H. Bunt, Phillip S. Jones, and Jack D. Bedient. Fundamental underpinnings of modern arithmetic, algebra, geometry, and number systems derived from ancient civilizations. 320pp. 5⅜ x 8½. 25563-8

A HISTORY OF MATHEMATICAL NOTATIONS, Florian Cajori. This classic study notes the first appearance of a mathematical symbol and its origin, the competition it encountered, its spread among writers in different countries, its rise to popularity, its eventual decline or ultimate survival. Original 1929 two-volume edition presented here in one volume. xxviii+820pp. 5⅜ x 8½. 67766-4

GAMES, GODS & GAMBLING: A History of Probability and Statistical Ideas, F. N. David. Episodes from the lives of Galileo, Fermat, Pascal, and others illustrate this fascinating account of the roots of mathematics. Features thought-provoking references to classics, archaeology, biography, poetry. 1962 edition. 304pp. 5⅜ x 8½. (Available in U.S. only.) 40023-9

OF MEN AND NUMBERS: The Story of the Great Mathematicians, Jane Muir. Fascinating accounts of the lives and accomplishments of history's greatest mathematical minds–Pythagoras, Descartes, Euler, Pascal, Cantor, many more. Anecdotal, illuminating. 30 diagrams. Bibliography. 256pp. 5⅜ x 8½. 28973-7

HISTORY OF MATHEMATICS, David E. Smith. Nontechnical survey from ancient Greece and Orient to late 19th century; evolution of arithmetic, geometry, trigonometry, calculating devices, algebra, the calculus. 362 illustrations. 1,355pp. 5⅜ x 8½. Two-vol. set. Vol. I: 20429-4 Vol. II: 20430-8

A CONCISE HISTORY OF MATHEMATICS, Dirk J. Struik. The best brief history of mathematics. Stresses origins and covers every major figure from ancient Near East to 19th century. 41 illustrations. 195pp. 5⅜ x 8½. 60255-9

Physics

OPTICAL RESONANCE AND TWO-LEVEL ATOMS, L. Allen and J. H. Eberly. Clear, comprehensive introduction to basic principles behind all quantum optical resonance phenomena. 53 illustrations. Preface. Index. 256pp. 5⅜ x 8½. 65533-4

QUANTUM THEORY, David Bohm. This advanced undergraduate-level text presents the quantum theory in terms of qualitative and imaginative concepts, followed by specific applications worked out in mathematical detail. Preface. Index. 655pp. 5⅜ x 8½. 65969-0

ATOMIC PHYSICS: 8th edition, Max Born. Nobel laureate's lucid treatment of kinetic theory of gases, elementary particles, nuclear atom, wave-corpuscles, atomic structure and spectral lines, much more. Over 40 appendices, bibliography. 495pp. 5⅜ x 8½. 65984-4

A SOPHISTICATE'S PRIMER OF RELATIVITY, P. W. Bridgman. Geared toward readers already acquainted with special relativity, this book transcends the view of theory as a working tool to answer natural questions: What is a frame of reference? What is a "law of nature"? What is the role of the "observer"? Extensive treatment, written in terms accessible to those without a scientific background. 1983 ed. xlviii+172pp. 5⅜ x 8½. 42549-5

AN INTRODUCTION TO HAMILTONIAN OPTICS, H. A. Buchdahl. Detailed account of the Hamiltonian treatment of aberration theory in geometrical optics. Many classes of optical systems defined in terms of the symmetries they possess. Problems with detailed solutions. 1970 edition. xv+360pp. 5⅜ x 8½. 67597-1

PRIMER OF QUANTUM MECHANICS, Marvin Chester. Introductory text examines the classical quantum bead on a track: its state and representations; operator eigenvalues; harmonic oscillator and bound bead in a symmetric force field; and bead in a spherical shell. Other topics include spin, matrices, and the structure of quantum mechanics; the simplest atom; indistinguishable particles; and stationary-state perturbation theory. 1992 ed. xiv+314pp. 6⅛ x 9¼. 42878-8

LECTURES ON QUANTUM MECHANICS, Paul A. M. Dirac. Four concise, brilliant lectures on mathematical methods in quantum mechanics from Nobel Prize–winning quantum pioneer build on idea of visualizing quantum theory through the use of classical mechanics. 96pp. 5⅜ x 8½. 41713-1

THIRTY YEARS THAT SHOOK PHYSICS: The Story of Quantum Theory, George Gamow. Lucid, accessible introduction to influential theory of energy and matter. Careful explanations of Dirac's anti-particles, Bohr's model of the atom, much more. 12 plates. Numerous drawings. 240pp. 5⅜ x 8½. 24895-X

ELECTRONIC STRUCTURE AND THE PROPERTIES OF SOLIDS: The Physics of the Chemical Bond, Walter A. Harrison. Innovative text offers basic understanding of the electronic structure of covalent and ionic solids, simple metals, transition metals and their compounds. Problems. 1980 edition. 582pp. 6⅛ x 9¼. 66021-4